T0330357

THE ECONOMICS OF AIR POLLUTION IN CHINA

THE ECONOMICS OF AIR POLLUTION IN CHINA

Achieving Better and Cleaner Growth

Ma Jun

Translated from the Chinese by Bernard Cleary
Edited in English by Damien Ma

A Paulson Institute Book

Columbia University Press
New York

Columbia University Press
Publishers Since 1893
New York   Chichester, West Sussex
cup.columbia.edu

保尔森基金会
PAULSON INSTITUTE

Columbia University Press gratefully acknowledges the support
of the Paulson Institute.

Library of Congress Cataloging-in-Publication Data
Names: Ma, Jun, 1964- author. | Ma, Damien, editor.
Title: The economics of air pollution in China : achieving better and
cleaner growth / Ma Jun ; translated from the Chinese by Bernard Cleary ;
edited in English by Damien Ma.
Other titles: PM2.5. English
Description: New York : Columbia University Press, [2017] |
Originally published in Beijing in 2014 as:
PM2.5 : jian pai de jing ji zheng ce. | Includes index.
Identifiers: LCCN 2016014649 | ISBN 9780231174947
(cloth : alk. paper)
Subjects: LCSH: Air pollution—Economic aspects—China. |
Economic development—Environmental aspects—China. |
Environmental policy—Economic aspects—China. |
Sustainable development—China.
Classification: LCC HC430.A4 M3213 2016 |
DDC 363.739/20951—dc23
LC record available at
https://lccn.loc.gov/2016014649

Columbia University Press books are printed on permanent
and durable acid-free paper.

Printed in the United States of America

Cover design: Adam Bohannon
Cover image: © Tom Nagy/Gallery Stock

Contents

Preface
vii

Introduction
1

PART ONE
Getting to 30 μg/m³

Introduction to Part One
13

CHAPTER ONE
PM2.5 Data, Reduction Model, and Policy Package
19

CHAPTER TWO
Environmental Actions: Necessary but Insufficient
40

CHAPTER THREE
Structural Adjustment: The *What* and the *How*
53

CHAPTER FOUR
Enabling Change: Incentives Needed
77

CHAPTER FIVE

The Cleanup and Economic Growth

101

PART TWO
Case Studies and Green Finance

CHAPTER SIX

Case Study: Shanghai

131

CHAPTER SEVEN

Case Study: Beijing

163

CHAPTER EIGHT

How to Deal with Coal

199

CHAPTER NINE

Making Green Finance Work in China

220

Notes

265

Index

285

Preface

AT AN OCTOBER 2013 meeting to discuss the results of this book, He Kebin, dean of the Tsinghua University School of Environment, said that "Most of our scholars and officials in the environmental field are engineers. Today, I'm very pleased to see a group of economists such as Dr. Ma Jun begin to focus their attention and expend their energy on the study of the pollution problem. This indicates the beginning of real hope for our environment."

The dean's statement highlighted a long-standing source of anxiety among environmental experts. They realized early on that environmental problems cannot be solved in the laboratory alone or by simply applying certain clean technologies or emissions standards. There have to be *economic* reasons that explain why emissions far exceed China's environmental capacity, and why the pollution problem worsened despite the introduction of many environmental regulations.

Yet for years, few have clearly articulated the economic rationale, and even fewer have quantified the relationship between structural factors in the economy and air pollution. Many know only very generally that environmental protection is an economic issue, but they do not know *how much* pollution is created by economic factors. And it naturally follows that not many have studied the degree of policy adjustment required to address these economic issues.

This book, based on the major findings of a research report titled "Economic Policies to Reduce PM2.5 Emissions," jointly supported by the Boyuan Foundation and Energy Foundation China, systematically quantifies the respective relationships between China's industry, energy, and transportation structures and air pollution. The intended focus of this project is to anchor the PM2.5 emissions reduction objective squarely in the changes necessary to alter China's economic structure. We argue that without a combination of policies and reforms to drive structural changes in the Chinese economy, the medium- to long-term air pollution goals will be very difficult, if not impossible, to achieve.

The book is divided into two parts. Part one (chapters 1–5) discusses economic policies to reduce PM2.5 emissions nationwide; quantitatively analyzes the effects of economic, energy, and transportation structures on air pollution levels; proposes economic policies to promote structural adjustments; and assesses the effectiveness of such policies. Part two (chapters 6–9) presents case studies of economic policies to reduce PM2.5 emissions in Shanghai and Beijing, discusses the economic policies to control consumption of conventional coal, and recommends the establishment of a green financial system in China.

The findings in the final chapter, on green finance, informed a green finance task force I led in mid-2014 at the People's Bank of China (PBOC). The task force aimed to systematically study how financial policies should be used to advance green investment and catalyze the green transformation of the Chinese economy. In collaboration with the United Nations Environment Programme's (UNEP) Inquiry into the Design of a Sustainable Financial System, the task force proposed fourteen recommendations for establishing China's green financial system, most of which were subsequently adopted in the Central Party Committee and State Council's "Comprehensive Plan for Institutional Reform to Promote Eco-Civilization" issued on September 21, 2015. Moreover, the development of a green financial system will also be incorporated into China's thirteenth Five-Year Plan for Financial Sector Development and Reform. The PBOC is now working with partners such as the Bank of England and UNEP to promote international cooperation on green finance under the G20 framework.

I led this book project with Li Zhiguo, a professor at Fudan University's business school. Other researchers who contributed include Shi Yu of Deutsche Bank; Xiao Mingzhi of Zhuoyue Development Research Institute; Chen Yuyu, professor of Applied Economics at Peking University; Zhang Yan, lecturer in the Department of Environmental Science at Fudan University; Yu Kun, doctoral candidate at the Fudan University Business School; and Zong Qingqing, postdoctoral research fellow in the Department of Applied Economics at Peking University. Graduate students Yu Kun, Xiao Xie, Quan Qiwei, and Li Xiangyu at Fudan University's business school and School of Environmental Science, and Lin Linzhi, Chou Xincheng, and Yan Anran at the Peking University Institute of Economic Policy Research also contributed research and other support.

The original Chinese language version of this work was published by China Economic Press in December 2014. With a total of fifteen chapters, the Chinese language version also included part three, led by Li Zhiguo, which compared international experiences of using policy options to mitigate air pollution, including taxes, clean energy incentives, automobile emissions controls, car license plate auction systems, and public transportation. Based on comments from the English publisher and external reviewers, and in light of the relative familiarity of readers abroad with international experience, part three is not included in the English edition of the book. Other discrepancies between the original Chinese language and English publications are primarily the result of editorial discretion, translation differences, and other considerations of target audience.

Authors of the various chapters in this English language version are as follows: introduction (Ma Jun), chapter 1 (Ma Jun, Shi Yu), chapter 2 (Ma Jun, Shi Yu), chapter 3 (Ma Jun, Shi Yu), chapter 4 (Ma Jun), chapter 5 (Ma Jun, Xiao Mingzhi), chapter 6 (Li Zhiguo, Ma Jun, Zhang Yan, Yu Kun), chapter 7 (Chen Yuyu, Ma Jun, Yan Se, Zong Qingqing), chapter 8 (Ma Jun, Xiao Mingzhi), and chapter 9 (Ma Jun, Shi Yu). I would like to thank all of my co-authors for their contribution and collaboration, and offer special thanks to Shi Yu, who co-authored several chapters and provided excellent data analyses and administrative support.

I'd also like to thank Columbia University Press and the Paulson Institute for their generous support of this project and dedication to making the English edition accessible to a wider audience. In particular, I want to thank Damien Ma of the Paulson Institute and Bridget Flannery-McCoy of Columbia University Press for their exceptional editorial skills and patience in bringing this project to fruition. Last but not least, I greatly appreciate Bernard Cleary's professional translation work.

THE ECONOMICS OF AIR POLLUTION IN CHINA

Introduction

FOR TWO DAYS in mid-January 2013, the PM2.5 air pollution index in Beijing briefly topped 900 and approached 1,000 micrograms per cubic meter ($\mu g/m^3$). This is almost forty times greater than the World Health Organization's (WHO) interim target-2 for the PM2.5 index of 25 $\mu g/m^3$—the standard target that all countries use for pollution reduction (see box on p. 2). The severe pollution effects were felt not only in Beijing but also in central and eastern China. According to the Ministry of Environmental Protection (MEP), since mid-January 2013, air pollution has endangered the health of over 600 million people in seventeen provinces and municipalities. Only one-fifth of Chinese cities met the country's own newly created air pollution standards.[1] Public concern about air quality is on the rise, and the health dangers of PM2.5 have generated broad anxiety. Numerous studies indicate the alarming health hazards of PM2.5 pollution.

Indeed, as knowledge about PM2.5 and its health effects became more widespread, the public concern has skyrocketed. Many Chinese took to the Internet to voice their anxiety, as reflected in comments such as, "If a decrease in life expectancy is the consequence of rapid superficial economic growth, then what is the point?" Such collective fear was manifest in consumer behavior. For instance, during the first month of 2013, sales of several brands of imported air purifiers surged two or threefold. Domestic firms, such as Beijing-based Suning Corporation, also benefited as it maintained 300 percent year-on-year sales growth for its air purifiers through February 2014.

Public opinion and consumer demand will force the government to take actions that may have once seemed unimaginable. It is more

WHO Air Pollution Standards These particulate matter (PM) levels are contained in the WHO's Air Quality Guidelines, which stipulate that annual average concentration of particulates having a diameter of 2.5 microns (PM2.5) are not to exceed 10 μg/m³. Because some countries and regions would need to take extraordinary steps to achieve the guideline requirements, the WHO has proposed three interim targets.

	PM10 (μg/m³)	PM2.5 (μg/m³)	Basis for selecting concentration
Air quality guidelines (AQG)	20	10	These are the lowest levels at which total cardiopulmonary and lung cancer mortality have been shown to increase with more than 95% confidence in response to long-term exposure to PM2.5.
Interim target-1	70	35	These levels are associated with about a 15% higher long-term mortality risk relative to the AQG level.
Interim target-2	50	25	In addition to other health benefits, these levels lower the risk of premature mortality by approximately 6% [2–11%] relative to the Interim target-1 level.
Interim target-3	30	15	In addition to other health benefits, these levels reduce the mortality risk by 6% [2–11%] relative to the Interim target-2 level.

Source: WHO, "Air Quality Guidelines—Global Update 2005."

than a pollution issue—if left unaddressed, social stability could be threatened. And this is why the central government has put significant political capital behind an economic reform program that is also meant to more vigorously tackle pollution and environmental challenges.

But beyond simply having the resolve to reform, the government also needs to determine the appropriate approach to such sweeping reforms. In early 2013, for example, the MEP announced that it would "strive for air quality in all cities nationwide to conform to the Stage Two Air Quality Standard by 2030," including a reduction in average annual PM2.5 to 35 µg/m³ from an estimated average of 65 µg/m³ (see chapter 1 for further explanation of how this estimate was derived).

However, considering the vast variations in pollution levels from city to city, to achieve the 35 µg/m³ target nationally by 2030 implies that the average level for Chinese cities must be reduced further, to 30 µg/m³. In other words, between 2013 and 2030, PM2.5 emissions should drop 54 percent (all subsequent projections and analyses are based on the 30 µg/m³).

Following the MEP proposal, the State Council also announced in September 2013 its "Action Plan for Air Pollution Prevention and Control," which explicitly requires that by 2017, average annual PM2.5 levels in the Beijing-Tianjin-Hebei region (also called "Jing-Jin-Ji") should be reduced by 25 percent, the Yangtze River Delta (YRD) region by 20 percent, and the Pearl River Delta (PRD) region by 15 percent.

This policy opened the floodgates to a slew of proposals and recommendations—more than eighty have been proposed in Beijing alone—from various scholars and those focused on environmental protection. The bulk of the measures have to do with end-of-pipe controls, such as desulfurization and denitration, increasing the quality of petroleum products, raising automobile emissions standards, and increasing fuel efficiency. Many of the proposed PM2.5 emissions reduction measures are administrative in nature, such as shutting down factories and construction sites and limiting the use of automobiles.

Our research indicates, however, that if China depends largely on end-of-pipe actions without addressing deeper structural economic problems—including excessive dependence on heavy industry, a low proportion of clean energy in total energy consumption, overreliance on road transportation, and the interregional externalities of environmental pollution—then only about half of the long-term national PM2.5 reduction targets, as well as those specifically for Beijing and

Shanghai, will be achieved. If the actions taken are predominantly administrative in nature, the economic impact will be significant in the short term, but will do little to drive structural adjustments.

Before delving into a detailed analysis of the structural issues in the Chinese economy that contribute to air pollution, a brief review of existing work on air pollution, and the proposed actions aimed at addressing it, is in order.

Literature Review

An enormous body of academic literature exists about air pollution and control in China. Content relevant to this book is primarily concentrated in three areas. The first concerns the severity of pollution and the economic costs of pollution. The second concerns analysis of the sources of atmospheric pollution and corresponding end-of-pipe control measures. The third concerns analysis of the deeper causes of pollution and control measures, including the economic structure, systems, and corresponding reform measures. The following summarizes this literature and highlights the new contributions made in this book.

I. On Health Hazards and Economic Losses

1. Studies conducted by Xie Peng and colleagues in Shanghai, Hong Kong, Beijing, Wuhan, and Taiyuan indicate that for every increase of 10 μg/m³ in atmospheric concentrations of PM2.5, the general acute mortality rate, mortality rate from cardiovascular disease, and mortality rate from respiratory disease increases 0.4 percent, 0.53 percent, and 1.43 percent, respectively.[2]

2. A joint study by Peking University and Greenpeace indicates that in the year 2010 alone, PM2.5 pollution in Beijing, Shanghai, Guangzhou, and Xi'an led to 7,770 premature deaths and economic losses totaling 6.17 billion yuan (~$1 billion).[3]

3. In a paper published in *Proceedings of the National Academy of Sciences of the United States of America*, Professor Chen Yuyu of Peking University, Professor Li Hongbin of the School of Economics and Management at Tsinghua University, and colleagues ingeniously used China's unique Huai River heating demarcation line to resolve the conundrum of accurately measuring causal relationships.

This paper concluded that life expectancies for the 500 million residents of northern China (north of Huai River) are reduced by an average of five and a half years due to heavy air pollution caused by coal burning.[4]

4. A report published by the US-based Institute for Health Metrics and Evaluation and Tsinghua University claimed that in 2010, outdoor air pollution in China led to 1.2 million premature deaths and to a reduction in healthy life-years for 2.5 million people.[5]

5. The Greater London Authority announced that a report by Dr. Brian G. Miller indicated a permanent increase in PM2.5 concentrations of 1 µg/m³ reduces resident life expectancy by three weeks.[6]

6. An analysis conducted by the American scholars Pope and colleagues, based on data from fifty-one American cities, indicated that if the PM2.5 concentration increases from 10 µg/m³ to 30 µg/m³, average life expectancy declines by two years.[7]

In addition to the health effects of air pollution, there is increasing concern over its economic cost. Air pollution can lead to economic consequences including increased healthcare expenses and declines in labor productivity. Projections based on existing research indicate that the annual losses brought about by air pollution may exceed 1 trillion yuan (~$160 billion) in China, calculated using 2013 prices. Several representative studies are highlighted below.

In a 2007 report titled "Cost of Pollution in China: Economic Estimates of Physical Damages,"[8] the World Bank systematically applied two methods of evaluating the value of statistical life (VSL): the adjusted human capital (AHC) method and the willingness-to-pay (WTP) method. Using the WTP method, the results demonstrated that for the year 2003, air pollution in China resulted in economic losses of approximately 520 billion yuan ($83 billion), equivalent to 2.8 percent of GDP. Even when the AHC method was used, the results showed losses as high as 157.3 billion yuan ($25 billion). Assuming the ratio of losses in 2013 remained at 2.8 percent of same-year GDP, the cost of pollution would be 1.6 trillion yuan ($258 billion), calculated using 2013 prices.

A 2010 study from the UK-based consultancy Trucost estimated the externalities of pollution in Asia according to the firm's own "Natural Capital Liabilities" method.[9] It found that economic losses caused by air pollution in East Asia in 2010 totaled roughly 930 billion yuan (~$150 billion). These economic losses included the impacts on health, crops, vegetation, buildings, and water. Because energy consumption in China accounts for approximately 75 percent of overall East Asian energy consumption, if assuming that pollution emissions are roughly proportional to energy consumption, then the losses caused by pollution in China totaled approximately 700 billion yuan (~$108 billion) in 2010. If it is further assumed that the ratio of pollution losses to GDP has remained stable, then the pollution losses calculated using 2013 prices totaled about 1 trillion yuan (~$161 billion).

Many Chinese and foreign experts have assessed in detail the severity of the pollution problem in China. As early as 2006, in the "Report on China's Green National Accounting Study,"[10] the State Environmental Protection Administration (SEPA, the precursor to the MEP) and the National Bureau of Statistics (NBS), using the pollution loss method, calculated the total cost of environmental pollution in 2004 to be 511.8 billion yuan ($62.4 billion), accounting for 3.1 percent of GDP. Disaggregating the total figure showed that the cost from air pollution was 219.8 billion yuan ($26.8 billion), water pollution 286.3 billion yuan ($34.9 billion), land use for solid waste 650 million yuan ($79.3 million), and contamination accidents 5.09 billion yuan ($621 million), accounting for 42.9 percent, 55.9 percent, 0.1 percent, and 1.1 percent of total environmental costs, respectively.

The World Bank was one of the first foreign institutions to attempt to determine the economic cost of pollution in China. Using the WTP method, it estimated that air pollution led to approximately 520 billion yuan (~$63 billion) in economic losses in 2003, or 3.8 percent of GDP. When water pollution is added, however, total environmental costs rose to 5.8 percent of GDP. In addition, in its 2014 study, "The Cost of Air Pollution,"[11] the Organisation for Economic Co-operation and Development (OECD) estimated that the economic costs of the health effects of China's air pollution were as high as $1.4 trillion.

While these studies certainly underscore the severity of environmental problems by analyzing the economic losses from pollution, they generally do not present systematic solutions.

II. Measures and Actions to Address Pollution

Countless research papers have been published on the sources of air pollution and the relevant end-of-pipe measures. For example, Chen Tianbing and colleagues (2006)[12] systematically probed atmospheric pollution caused by coal consumption, as well as coal pollution control technologies such as desulfurization before combustion, sulfur retention during combustion, purification after combustion, establishment of large pithead power plants that utilize the dry coal-cleaning method, and developing coal conversion technology. In another study, Chen Danjiang and Yang Guang (2013)[13] saw coal as the primary source of air pollution and proposed implementing desulfurization and denitration for cleaner production and enhancing treatment of wastewater and exhaust emissions to render them innocuous, based on circular economy principles.

As early as 2001, the Environmental Science and Engineering Department at Tsinghua University and SEPA put forward a "China Motor Vehicle Emissions Control" plan,[14] providing detailed estimates of the environmental benefits that can be gained from upgrading petroleum products and switching from gasoline to natural gas. Ruan Xiaodong (2013)[15] found that the poor quality of China's petroleum products, and in particular their relatively high sulfur content, was directly linked to the current atmospheric pollution problem, though technical difficulties remain in terms of upgrading those products. A study by Zhang Chanjuan and Chen Xiaojun (2013)[16] also found that motor vehicle exhaust emissions were a primary factor in the formation of smog, and proposed relevant motor vehicle administration and exhaust control measures.

Environmental protection agencies, too, have compiled and issued relevant policies and technical guidelines to mitigate air pollution, including *Technical Specifications for Managing the Operation of Flue Gas Treatment Facilities at Thermal Power Plants* (2012),[17] *Guide to Best Practical Technologies for Pollution Prevention in the Cement Industry* (2012),[18] *Technical Policy for the Prevention of Volatile Organic Compound* (VOC) *Pollution* (2013),[19] and *National China V Fuel Economy Standard for Vehicles* (2013).[20]

III. Economic Causes of Pollution

As the air pollution problem has become more urgent, an increasing number of scholars have recognized the need to study and analyze air pollution control measures from an economic perspective. In a 2012 paper titled "A High Level of Importance Must be Attached to the PM2.5 Pollution Prevention and Control Problem," Zhou Hongchun of the State Council Development Research Center[21] pointed out that imperfect regulatory standards, incomplete statistical and review indicators, and the lack of mechanisms to establish regional linkages are all fundamental causes of severe PM2.5 pollution in China. The study further recommended short-term actions, such as strengthening monitoring, controlling pollution from construction projects, and scrapping "yellow-label vehicles" (essentially those that do not meet fuel standards), as well as long-term control measures such as optimizing the industrial and energy structure, utilizing market mechanisms to control pollution, strengthening regional linkages, and creating enduring PM2.5 pollution control mechanisms.

Lu Shize (2012)[22] at the MEP also proposed four major measures: "centralize planning of regional environmental resources and optimize the structure and configuration of industry; strengthen the utilization of clean energy and control regional coal consumption; intensify the total volume of atmospheric pollution control efforts and implement synergistic controls of multiple pollutants; and create regional management mechanisms and improve joint prevention and control management capabilities."

Approaching the issue from an economic perspective, Liu Zixi (2013)[23] proposed five major strategies to control air pollution: improving the energy production and consumption structure and deploying clean energy; optimizing resource allocation and achieving balanced regional development; accelerating optimization and upgrading of the industrial structure and promoting development of tertiary industries; standardizing approval and oversight of industrial projects and pressing ahead with energy conservation and emissions reduction efforts; and strengthening basic industries and infrastructure construction and developing public transportation.

In addition, Han Wenke and colleagues (2013)[24] argued that the primary causes of smog were the long-standing irrational urban economic and energy structures, ineffective motor vehicle emissions controls, inadequate atmospheric pollution prevention and control measures, and the severe imbalance in the development of urban periphery areas. Therefore, to transform the economic development model and optimize the energy structure, China must substantially increase the proportion of clean energy used in cities and reduce motor vehicle tailpipe emissions by developing public transportation while improving the quality of petroleum products. At the same time, joint prevention and control measures are essential.

Finally, Lin Boqiang (2013)[25] has given his expert views on the energy structure and the development of new energy, while Jia Kang (2013)[26] has proposed a pollution control scheme based on a fiscal policy approach.

International scholars have also conducted extensive analyses of China's air pollution challenges. Relatively well-known among these are *Clearing the Air* and *Clearer Skies Over China: Reconciling Air Pollution, Climate, and Economic Goals*, jointly written by Chris P. Nielsen and Mun S. Ho in 2007 and 2013,[27] respectively. In their 2013 work, the authors discussed the current state of China's air pollution and policies, estimated pollution emissions of coal-burning power plants and the cement industry, used a model to quantify the pollution inventory and concentrations as well as the economic benefits of cutting emissions, and presented a series of policy recommendations.

Many of the studies discussed above have played a positive role in pushing the Chinese government toward the creation of an atmospheric pollution control program. Overall, however, the package of policies currently in place continues to focus on end-of-pipe controls and largely relies on administrative measures. This is because both the research on economic policies and the policies' implementation to control smog have been inadequate.

The studies currently available on the economic causes of air pollution and the relevant economic policies are flawed in the following respects:

1. The majority of the studies focus on a qualitative approach and fail to quantify the respective roles played by the current industry, energy, and transportation structures in the formation of smog. They also do not quantify the extent to which PM2.5 can be reduced by changing these structural distortions.

2. The majority of the studies state that the economic structure needs adjustment, but they offer no in-depth analysis of specific policies to drive such structural adjustment. They also do not specifically quantify the intensity of such measures and implementation costs—for example, their effects on economic growth, fiscal revenue, and inflation.

3. The majority of existing studies focus on the macro level and do not consider the many problems local governments would encounter at the micro, implementation level—issues such as financing, regional ecological compensation mechanisms, and automobile license plate controls.

4. Most of the economic studies focus on controlling pollution using fiscal and tax policy, but very few propose market-based financing solutions, specifically a policy framework to catalyze green financing in China.

Compared to the existing literature, our approach in this book distinguishes itself in several respects. First, we have developed a PM2.5 control model that quantifies the effects of major end-of-pipe control measures and the various structural changes (industry, energy, and transportation) on the reduction of PM2.5 emissions. We propose that it will be necessary to rely on structural adjustments to achieve approximately half of the proposed emissions reduction target. Second, this work systematically proposes ten specific policy measures to drive structural adjustments, and quantifies the intensity and implementation costs of these measures. Third, this work incorporates a detailed discussion of the economic policies, as well as practical policies and measures, that local governments can use to deal with smog, focusing on the specific cases of Beijing and Shanghai. Fourth, this book provides a quantitative analysis of the effectiveness of the policy package to control conventional coal use. Finally, the book provides a foundation for the establishment of a green financing policy framework, as well as suggested measures.

PART ONE

Getting to 30 μg/m³

Introduction to Part One

In our research, we found that the existing economic, energy, and transportation structures in China (hereinafter referred to as the "economic structure") are major contributors to PM2.5 pollution levels. To meet the Chinese government's air pollution reduction target of 35 μg/m³ for all cities, we estimate that by 2030, the average urban PM2.5 level should drop to 30 μg/m³. The quantitative analysis that follows is based on the objective of meeting this target. Before discussing how and why the economic structure settled into its current state, it is worthwhile to summarize what this current structure looks like.

I. Heavy Industry Has Grown Too Quickly

Industry as a proportion of China's GDP is 40 percent, and heavy industry as a proportion of total industrial output from enterprises above a designated size—those with annual turnover above 20 million yuan ($3.2 million)[1]—is as much as 72 percent (based on 2011 data).[2] In other words, heavy industry accounts for as much as 30 percent of GDP, a proportion that is the highest among major global economies. In the last dozen years or so, the central government has consistently emphasized the need for structural adjustment, or "rebalancing," to increase consumption as a proportion of GDP, yet investment has in fact continued to increase. This means that energy-intensive and highly polluting heavy industry has also risen as a proportion of GDP.

For example, between 2000 and 2011, heavy industry as a proportion of total industrial output from industrial enterprises above a designated size rose from 60 percent to 72 percent. Unfortunately, energy intensity in industry—or the energy consumed per unit of output—is four times that of the services sector (for heavy industries, it is nine times that of the services sector). Even if the average annual growth rate of real output from heavy industries were to decelerate from 13 percent (in the decade 2003–2012) to around 6.8 percent from 2013 to 2030, real output from heavy industry would still grow a cumulative 206 percent during this period.

Under this scenario, even if pollution emissions per unit of output were reduced by as much as 80 percent (technically difficult to achieve), it would nonetheless be impossible to realize the 54 percent reduction in PM2.5 emissions needed to achieve the 30 µg/m³ target.

II. Rapid Growth in Coal Consumption

Coal consumption is the most significant source of PM2.5 emissions in China. However, under current policies that are very friendly to the coal industry—including low resource taxes and extremely low pollution emissions fees—consumption of this commodity has consistently exceeded projections for many years. In fact, coal's share in overall energy consumption has remained virtually unchanged over the last thirty-five years, at around 70 percent.

Even if the average annual coal consumption growth were to decline from the 8 percent seen in the five years from 2007 to 2012, to 4 percent over the following decade (2013–2022), China's annual coal consumption would still increase from around 3.8 billion tons to 5.6 billion tons in 2022. This demonstrates that existing policies have not been effective in curbing the pace of growth in coal consumption. Without major policy changes, it will be impossible to reduce coal's proportion in total energy consumption as planned.

III. Passenger Vehicle Use Increasing, Rail Use Lagging

Automobile tailpipe emissions are the second-largest source of PM2.5 emissions in China. In the years 2007–2012, China's passenger vehicle ownership rose on average 20 percent per year. In Beijing, one of the most heavily polluted cities, passenger vehicle ownership per thousand people has now reached 190 vehicles, far higher than Singapore's 110 (see later chapters on discussion of the Singapore model). Based on current trends and policies, some industry experts forecast that the number of passenger vehicles in China will increase from the current 90 million to 350–400 million in 2030, averaging roughly 8 percent annual growth. They argue that 400 million passenger vehicles translates to a penetration rate of just 280 vehicles per 1,000 people, far lower than the 40–50 percent found in Europe and 63 percent in the United States.[3] Moreover, China's current ambitious urbanization strategy may continue to incentivize high growth in car ownership. That's because the emphasis is on developing small and medium-sized cities while curtailing the growth of the largest cities. Since smaller cities are generally unlikely to have dense urban public transportation networks, such as subways, the current urbanization approach could encourage a car-intensive development model.

For the same unit of traffic volume, the volume of PM2.5 emissions from subway and rail transport is less than one-tenth the volume of emissions from road transport. Development of railroads and subways is an important step toward curbing air pollution. The government's current plan is to increase the total railway network from 90,000 kilometers (km) in 2011 to 120,000 km in 2015, and experts project that total railway length could be extended to nearly 140,000 km by 2020. Likewise, the central government plans to expand the total length of subways from the current 2,000 km to 7,000 km by 2020.[4]

Although these long-term objectives appear impressive on paper, they actually constitute a "negative contribution" to PM2.5 emissions reduction because these growth rates are still too low, according to our projections. We estimate that the current railway and subway expansion plans imply an average of less than 4 percent growth through 2020 in terms of rail and subway traffic volume, far lower than the 6.1 percent average growth projected for total traffic volume nationwide over the same period. If the lower growth in railroads and subways persists, then the number of passenger vehicles will continue to average 8 percent annual growth. This implies that the total number of vehicles will reach 360 million by 2030, four times the current level.[5]

Of course, in the meantime, the Chinese government plans to improve fuel quality, automobile emissions standards, and fuel efficiency, while implementing stricter emissions standards for thermal power generation and industrial coal consumption. But none of these measures will materially alter the economic structure.

Given the current economic structure, let us now pose three assumptions: (1) China's industry and heavy industry as proportions of GDP do not change; (2) coal consumption as a proportion of total energy consumption does not change; and (3) the number of passenger vehicles will increase by an additional 300 percent, as experts predict. Based on these assumptions, and assuming that unit coal consumption and unit automobile emissions do not change, our study demonstrates that even if maximum efforts are employed to deploy clean coal technologies (desulfurization and denitration), improve petroleum product quality, enhance automobile emissions standards, and increase fuel efficiency, China's average annual PM2.5 level will still be around 46 µg/m³ by 2030. In other words, if China wants to reduce PM2.5 to the targeted 30 µg/m³, it is imperative that additional efforts be pursued to significantly alter the economic structure.

Underlying Causes of Structural Distortions

How did the current economic structure develop? From an economic perspective, it is important to examine the roles of four players: the government, the market, businesses, and consumers. At least three factors have led to an energy- and pollution-intensive economic structure: government distortions of the market, market failure, and lack of a sense of social responsibility among businesses and consumers.

In China, routine government interventions that distort market functions have resulted in incentives that promote the extensive expansion of polluting industries. Yet due to the externalities of pollution, it is not possible to endogenize these externalities by relying solely on market mechanisms. This type of market failure has led to a combination of investment, production, and consumption behaviors that have resulted in significant pollution. Though it has failed to do so to date, the Chinese government would do well to design a variety of mechanisms that endogenize these pollution externalities to correct for market failures. Put another way, the government has intervened in the market where it should not and has not intervened in areas where the market has failed.

In general, firms seek to maximize profits and consumers seek to maximize consumption enjoyment. But if the objective functions of firms and consumers also include a dislike for pollution, or conversely, a preference for "green" products, then at the same market price, production and consumption of polluting products will fall, while production and consumption of green products will correspondingly rise. But one of China's key problems is a lack of concern for environmental protection among many firms and institutional investors, insufficient transparency in environmental information, inadequate environmental protection education, and weak public support for a green economy. Consumers and businesses do not feel a sense of shame when purchasing or producing polluting products, nor do they feel a sense of pride when using or producing green products.

Each of these problems, anchored in economics, will be examined in detail, and a more systematic framework will be provided for the proposed policies to reduce PM2.5 emissions.

Organization of Part One

Chapter 1 provides a detailed description of our quantitative model, which evaluates the effects of environmental and economic policies on PM2.5 emissions and is used to develop the package of policies required to achieve the PM2.5 reduction target. We discuss the model framework, the steps we took to arrive at our estimate, the source data (such as PM2.5 sources, current emissions levels, reduction targets, and the various elasticity coefficients), and the scope of the model's application.

Based on this model, in chapter 2 we discuss the results and our conclusion that if the current economic, energy, and transportation structures remain unchanged—even if efforts in the areas of clean coal technology, emissions standards, and policies achieve their maximum potential—the 30 µg/m³ target will still not be met.

In chapter 3 we analyze the three main structural issues that make thorough reduction of air pollution difficult to achieve—expansive heavy industry, reliance on coal consumption, and road transportation. Consequently, to meet the national PM2.5 target, China must undertake a series of aggressive structural adjustments, which are detailed as a proposed package of recommendations.

In chapter 4 we discuss the economic policies that must be adopted to facilitate the structural adjustments, while in chapter 5 we examine the likely economic impact of such policies. A key finding is that these policy measures will not lead to a substantial deceleration in economic growth from 2013 to 2030.

Part two includes extensive case studies of Beijing and Shanghai and the associated major policy recommendations for those cities to achieve their air pollution reduction targets. Finally, we conclude with chapters on analyzing policies to reduce coal consumption and establishing a green finance policy framework.

CHAPTER ONE

PM2.5 Data, Reduction Model, and Policy Package

IMPOSING TOUGHER ENVIRONMENTAL standards and intensifying their enforcement will undoubtedly result in reduced PM2.5 pollution. But such policies will not necessarily achieve the Chinese government's objective, since their effectiveness may well be offset by other economic, industrial, and fiscal policies that run counter to them. The appropriate policy package for reducing PM2.5 emissions requires a top-down design, and the key to this design lies primarily in the establishment of quantitative relationships between the impacts of environmental and economic policies on PM2.5 levels.

One of the key findings in our study is that if the current economic structure remains unchanged, even if efforts in deploying environmental technology and enforcing environmental standards achieve their maximum results, it will *still* be impossible to meet the PM2.5 emissions reduction target. Specifically, proposed environmental policies will together reduce the national urban PM2.5 level by only a little more than half, to 46 µg/m³ (from an average of 65 µg/m³), falling far short of achieving the 30 µg/m³ target.

These proposed environmental policies include cutting emissions per unit of coal by 70 percent through desulfurization and denitration, reducing unit automotive emissions per vehicle-km by 80 percent by improving fuel quality, increasing automobile emissions standards and fuel efficiency, and controlling volatile organic compounds (VOCs) and construction dust to reduce unit emissions generated by industry and construction by 80 percent. It would be nearly impossible to achieve the target by relying solely on such environmental measures, primarily due to major structural issues in the economy, as discussed above.

First, China's heavy industry as a proportion of GDP is the highest among large economies in the world, and the intensity of coal consumption in heavy industry is nine times that of the services sector. Second, clean energy accounts for only 14.6 percent of China's energy consumption (the average in OECD countries is 42 percent),[1] while coal accounts for 67 percent. Air pollution generated from burning conventional coal is ten times higher than from cleaner sources, given the same energy equivalent. Third, road transport accounts for more than 90 percent of urban commuting (the average in major cities of developed countries is 30 percent)[2] while the air pollution generated per passenger-km by car is ten times that of subway travel.

The contradiction between policies that incentivize the current economic structure on the one hand, and the stated environmental objectives on the other, are the fundamental reasons for long-standing difficulties in curbing air pollution. This fundamental contradiction reflects many deeper problems.

First, because the Chinese public has lacked sufficient awareness of the health risks posed by air pollution, political pressure to address the issue has been weak. As a result, the pollution problem has not been fully integrated into the top-down policy design process. Second, at the operational level, the central government has been subject to a form of "bureaucratic capture." Economic policies and strategy have often been based on plans submitted by various ministries. The bureaucracies often seem to represent the disparate interests of industries, and there is no single, powerful agency that has devoted attention to analyzing economic policies and their environmental implications. Third, in the area of policy research, no one to our knowledge has quantified the environmental impacts of various existing economic and industry policies, nor translated emissions reduction targets into requirements for changing the goals of sectoral development.

At present, the majority of air pollution control proposals primarily involve quantifying the effects of individual clean technologies—for example, determining the extent to which emissions can be reduced per kilowatt (kW) of electricity generation via desulfurization or the extent to which emissions/km can be reduced by improving the quality of petroleum products.

This study, then, attempts to bridge these gaps. We have built a "PM2.5 emissions reduction model" that simulates the effects of various policy packages on air pollution control. The remainder of the chapter discusses in detail the process by which our estimates are derived, the underlying data of this model, and the scope of its application.

I. The Data

1. Estimating PM2.5 Levels

The central government has yet to publish the average urban PM2.5 level (i.e., the mean of average annual concentrations in all cities for which data are collected) for 2012, the base year for this study. Without data for the base year, it is impossible to know the extent to which emissions need to be reduced going forward and impossible to estimate the intensity of the required policies. In the absence of official data on PM2.5 concentrations, we examined five available data sources as proxies to arrive at our own estimate of the average urban PM2.5 level in recent years.

First, we used Greenpeace's PM10 data for the years 2010–2012, which were calculated based on population-weighted averages in 120 cities nationwide. We used this data because the ratio of PM2.5 to PM10 in the same location generally ranges from 0.65 to 0.9, according to numerous academic studies.[3] Therefore, based on an average annual PM10 level of 92.7, and using a ratio coefficient of 0.7, we estimated urban average PM2.5 emissions to be 65 μg/m³.

Second, we determined the arithmetic mean of the PM2.5 benchmark for 71 major cities in China to be 71 μg/m³, based on 2013 data from the National Air Quality Software Company and the PM2D5 website.[4]

Third, using historical data from MEP for Tianjin, Shanghai, Chongqing, Suzhou, Guangzhou, Nanjing and Ningbo, we found the average PM2.5 concentration for these cities was 53 μg/m³ in 2010.[5]

Fourth, data collected by the National Aeronautics and Space Administration (NASA) Global Satellite Monitoring instruments indicated that in the five-year period from 2001 to 2006, average PM2.5 concentrations in major cities in eastern and central China were between 50 and 80 μg/m³. We took 65 μg/m³ as the average.[6]

Finally, in a paper published in *Environmental Science & Technology*,[7] Michael Brauer and his team used a combination of satellite, ground monitoring, and model data to estimate the average population-weighted annual PM2.5 concentration in the East Asia region, resulting in 55 µg/m³. Based on this estimate, as well as the population and pollution levels in various East Asian countries at the time, the PM2.5 level we derived for China was approximately 63 µg/m³.

All of these estimates are useful proxies, which we then averaged to arrive at a PM2.5 level of 65 µg/m³. This figure is our assumption for China's urban average PM2.5 level in 2012, the base year of our PM2.5 control model. It is worth noting that at the time of this writing, PM2.5 data for 2014 and the first nine months of 2015 became available.[8] Based on this more recent data for 74 cities, the urban average PM2.5 level for 2014 was about 64 µg/m³ and was about 58.5 µg/m³ for the first nine months of 2015. The three-year average from 2013 to 2015 is roughly 64.5 µg/m³, largely in line with our estimate.[9]

2. PM2.5 Data for Chinese Cities

PM2.5 levels also showed regional variation across urban China, as shown in table 1.1. Generally speaking, cities in northern China have higher PM2.5 levels than southern cities. One of the reasons for this difference is significant burning of coal for heat during northern China's winters. Moreover, major steel manufacturers are concentrated in the Beijing-Tianjin-Hebei (Jing-Jin-Ji) region, which means the density of emissions from coal burning is higher in this region than in most areas in the south. Yet another reason Beijing's PM2.5 levels are notably higher is that the Chinese capital has the highest number of cars among major Chinese cities. With private automobile ownership (refers to automobiles owned by individuals) totaling 4.1 million and civilian automobile ownership (includes automobiles owned and used by individuals, government agencies and enterprises, but excludes vehicles used for military, police, and other special purposes) totaling nearly 5 million in 2012, Beijing has two to three times the number of vehicles as Shanghai, which had a total of 1.4 million private and 2.1 million civilian vehicles as of 2012.[10]

3. Primary Sources of PM2.5

To effectively reduce PM2.5 emissions, we must first understand the sources and composition of the pollution. Our estimates, based on

TABLE 1.1
PM2.5 Levels in 71 Chinese Cities in 2013 (µg/m³)

Ranking	City	PM2.5	Ranking	City	PM2.5	Ranking	City	PM2.5
1	**Shijiazhuang**	149	26	Huzhou	75	51	Taizhou	54
2	Handan	132	27	Suqian	74	52	Ningbo	54
3	Baoding	130	28	Zhenjiang	73	53	Dalian	53
4	Hengshui	121	29	Nantong	72	54	Foshan	52
5	**Jinan**	117	30	Jiaxing	71	55	**Guangzhou**	52
6	Tangshan	116	31	**Changchun**	71	56	Chengde	52
7	Langfang	115	32	Suzhou	71	57	**Guiyang**	51
8	**Zhengzhou**	110	33	Qinhuangdao	70	58	Lishui	49
9	Xi'an	106	34	Shaoxing	70	59	Jiangmen	49
10	**Tianjin**	98	35	Yangzhou	69	60	Zhongshan	48
11	**Wuhan**	92	36	**Nanchang**	69	61	Dongguan	47
12	**Chengdu**	89	37	Jinhua	69	62	**Yinchuan**	47
13	**Beijing**	88	38	Lianyungang	68	63	Zhangjiakou	43
14	Urumqi	88	39	Yancheng	68	64	**Shenzhen**	40
15	**Hefei**	87	40	Qingdao	68	65	**Kunming**	39
16	**Changsha**	82	41	**Lanzhou**	68	66	Zhuhai	39
17	Huaian	81	42	**Xining**	67	67	Huizhou	37
18	Taizhou	80	43	**Hangzhou**	67	68	**Fuzhou**	34
19	**Taiyuan**	79	44	Quzhou	67	69	Xiamen	33
20	Harbin	78	45	**Chongqing**	65	70	**Lhasa**	27
21	**Nanjing**	77	46	**Shanghai**	62	71	**Haikou**	26
22	Xuzhou	77	47	**Hohhot**	61			
23	Wuxi	76	48	Wenzhou	57			
24	**Shenyang**	76	49	**Nanning**	56			
25	Changzhou	75	50	**Zhaoqing**	54			

Source: China Air Quality Software Company, www.pm2d5.com.

Note: Cities in bold type denote all provincial capitals and municipalities, as well as the Shenzhen Special Economic Zone, under the direct jurisdiction of the central government.

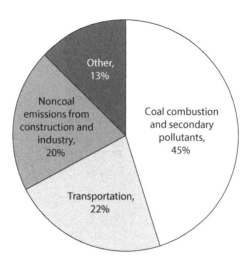

Figure 1.1 Composition of PM2.5 sources. *Source:* Author's estimate based on existing literature, including Greenpeace and Peking University studies.

existing research, indicate several major sources of PM2.5 pollution (see fig. 1.1). We found that nearly half of the PM2.5 emissions are from coal consumption and secondary sulfides and nitrogen oxides, roughly a quarter from transportation, about one-fifth from industry and noncoal burning construction, and the rest from other sources (including straw burning, chemical fertilizers, pesticides, tobacco smoking, cooking, forests, and oceans). The quantitative model uses these percentages as important baseline data.

Our estimates also draw on several existing academic works. For example, Greenpeace provided data that indicate about 49 percent of PM2.5 emissions in China come from coal burning and 16 percent from motor vehicle fuel combustion (see table 1.2). A study by Hu Min, director of the State Key Joint Laboratory of Environmental Simulation and the Pollution Control Laboratory at Peking University, indicates that the burning of fossil fuels, including coal and gasoline, accounts for 60–70 percent of PM2.5 emissions.[11] These percentages align with our estimate that coal and transportation together account for about 67 percent of PM2.5 emissions.

When categorized by industry (excluding transportation), three industries comprise the bulk of air pollutants: thermal power, cement, and steel (see figs. 1.2–1.4). The figures show industrial sources of

TABLE 1.2
Sources of PM2.5 Pollution

Thermal power plant coal combustion*	17%
Industrial coal combustion*	19%
Other industrial pollution	17%
Motor vehicle fuel combustion	16%
Urban residential and commercial coal combustion*	4%
Combustion of other fuels by urban residents and businesses	16%
Rural coal combustion*	8%
Rural biomass energy	1%
Rural combustion of other fuels	2%
Total	**100%**
***Proportion related to coal (rounded up)**	**49%**

Source: Greenpeace.

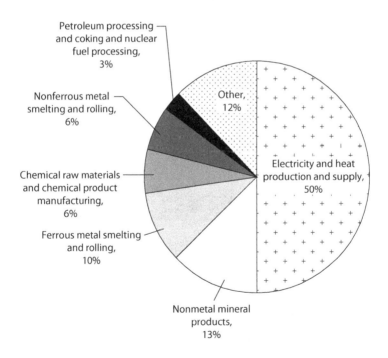

Figure 1.2 Composition of industrial sources of SO_2. *Source:* First National Pollution Survey 2007, NBS.

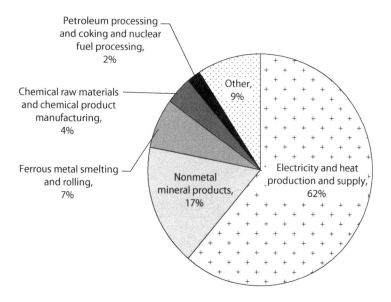

Figure 1.3 Composition of industrial sources of NOx. *Source:* First National Pollution Survey 2007, NBS.

Note: numbers do not add to 100 because of rounding.

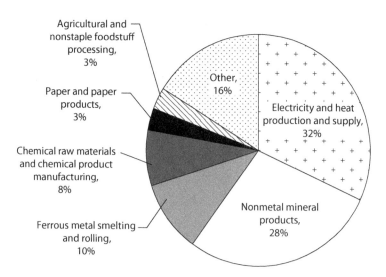

Figure 1.4 Composition of industrial sources of dust. *Source:* First National Pollution Survey 2007, NBS.

sulfur dioxide (SO_2), nitrous oxide (NOx), and dust, respectively, all of which contribute to PM2.5 emissions directly or indirectly.

No authoritative estimate exists for the composition of the sources of noncoal emissions from industry and construction, but multiple studies have suggested that the primary sources are industrial soot and dust, construction dust, and VOCs. Soot and dust emitted by industries and dust from construction sites and roads are primarily direct contributors of PM2.5, while VOCs become PM2.5 largely as secondary pollutants via chemical reactions. Studies of urban pollution have shown that construction dust accounted for 23 percent of the city of Nanchang's PM2.5 between 2007 and 2008;[12] urban dust accounted for 20.4 percent of Ningbo's PM2.5 in 2010;[13] metallurgical sources (noncoal) accounted for 5 percent of Jinan's PM2.5 in 2006;[14] and industrial emissions accounted for 16.4 percent of PM2.5 around Changsha.[15]

4. International Comparison of PM2.5 Levels

How serious is China's PM2.5 pollution compared to that of other countries? According to NASA's satellite data for 2001–2006, China's PM2.5 pollution is among the worst in the world—approaching or exceeding 80 $\mu g/m^3$—with the majority of the pollution in the northern, eastern, and central regions, which also have the densest urban populations. In contrast, PM2.5 levels in most areas in North America, Australia, and Russia are below 15 $\mu g/m^3$ (see fig. 1.5).

Taking a closer look, figure 1.6 compares PM2.5 levels in select Chinese cities with PM2.5 levels in major cities in developed countries. Note that average PM2.5 levels exceeded 100 $\mu g/m^3$ across many locales in China, including Shijiazhuang, Jinan, and Xi'an. Even in Haikou, the Chinese city with the best air quality, the average annual PM2.5 concentration is 26 $\mu g/m^3$, which exceeds that of all foreign cities in the graph. In contrast, PM2.5 levels in major European cities lie within the range of approximately 15–20 $\mu g/m^3$, while levels in most North American cities are as low as 10–15 $\mu g/m^3$.

II. The Model

Before delving into our model in detail, it is worth reiterating that we have proposed reducing the average urban PM2.5 concentrations from

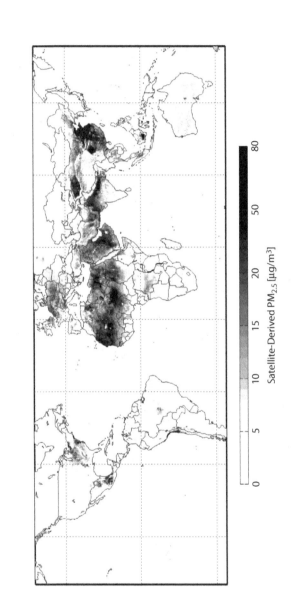

Figure 1.5 Global PM2.5 emissions intensity, 2001–2006. *Source:* NASA.

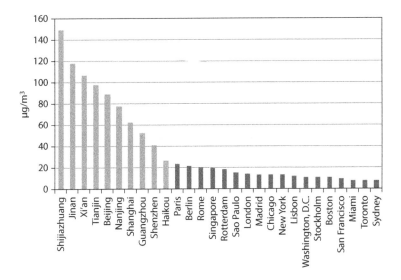

Figure 1.6 PM2.5 levels in various global cities (μg/m³). *Sources:* National Air Quality Software Company (www.pm2d5.com); WHO.

Note: Data for Chinese cities are average levels for 2013; figures for other cities are averages in recent years.The website from which these data were taken is no longer available. Similar data are now available on http://www.pm25.in/.

the current 65μg/m³ to 30 μg/m³ by 2030. This is because our estimates show that hitting this target is necessary in order for the majority of Chinese cities to realize MEP's official national PM2.5 target of 35 μg/m³ by the same year.

To that end, we built a "PM2.5 emissions reduction model" that quantifies the emissions reduction impact of various environmental measures and structural economic adjustments. Based on the results, we also propose the degree of environmental measures, as well as the extent of the structural changes to the industrial, energy, and transportation sectors, required to achieve the 2030 target.

1. Settling on the 30 μg/m³ Target for 2030

First, let us explain how we determined a target that is slightly lower than the official national target. Though the MEP announced in January 2013 a pledge to achieve the stage-2 air quality standard by reducing the annual average PM2.5 level to 35 μg/m³ in all cities by 2030, this official pledge means that the *average* PM2.5 actually needs to be reduced further.

This is because, in our view, "conformance to the standard for PM2.5 in all (i.e., each of the) cities" and "an average PM2.5 level for all cities that conforms to the standard" are two very different concepts. Since the pollution levels across cities vary widely, even if the average value for cities nationwide conforms to the standard, nearly half of the cities will nonetheless continue to exceed the standard. In other words, the requirement of "conformance to the standard by each city" is significantly higher than the requirement of "conformance to the standard by the average for all cities."

Therefore, based on the dispersion coefficient for PM2.5 (0.42) among 71 cities in the first quarter of 2013, and assuming this coefficient will decrease by half in the future, we found that in 90 percent of the cities, to reduce the PM2.5 level to 35 µg/m³ or less, the *average* PM2.5 level for these cities must be reduced to 30 µg/m³.

We set an urban average PM2.5 target lower than the official one for three additional reasons. First, the Chinese public has noted that the WHO's stage-2 interim PM2.5 target is only 25 µg/m³, and the public has questioned why the Chinese people should continue to tolerate pollution that is worse than the international standard. Second, the WHO's target is only an interim one, since the organization's Air Quality Guidelines stipulate that ultimately, the PM2.5 level should not exceed 10 µg/m³.[16] In other words, China's target of 35 µg/m³ is still 250 percent higher than the level WHO considers safe. Moreover, high-profile and influential voices, including Chinese People's Political Consultative Conference spokesperson Lü Xinhua, have expressed that 2030 is too long a wait for the Chinese public to breathe cleaner air. Thus, if public opinion is to be respected, PM2.5 reduction should err on the aggressive side. For all of these reasons, we use 30 µg/m³ as the target for our subsequent analysis.

2. Creating the Model

The PM2.5 emissions reduction model we developed establishes quantitative relationships between PM2.5 and a series of environmental measures such as end-of-pipe control and structural adjustments, including secondary industry as a proportion of GDP, coal consumption as a proportion of total energy consumption, and the ratio of road traffic to total traffic (see fig. 1.7). Based on this model, we recommend a package of actions and policies that will facilitate achieving the PM2.5 reduction target.

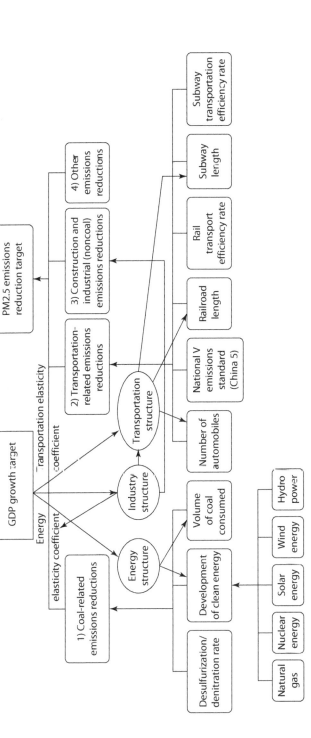

Figure 1.7 PM2.5 emissions control model. *Source:* Author.

It should be noted, however, that once the emissions intensity is controlled, this model assumes linear relationships between each of the primary sources of PM2.5 emissions (e.g., coal, automobiles, industrial noncoal emissions) and the primary economic variables (e.g., amount of coal burned, number of automobiles in use, real industrial output). This assumption does not apply to short-term PM2.5 forecasts and regional comparisons of PM2.5. That is because during different seasons of the year, and even during different times of the day, variable wind speeds can cause the same economic activity to produce wildly different PM2.5 concentrations. Climate conditions in the various regions can also result in very different pollution conditions. The intent of this model is to evaluate the long-term changes in the economic structure and emissions intensity and their impact through 2030; therefore we do not account for the effects of annual cyclical seasonal changes and regional differences in climatic conditions.

This quantitative analysis applies to the "baseline scenario" of implementing environmental end-of-pipe measures with no structural adjustment, and the more aggressive "structural adjustment scenario," which combines end-of-pipe actions with structural adjustment. We also apply this modeling exercise to a third but unlikely scenario that assumes nothing will be done—that is, neither environmental measures nor structural adjustment will take place (see chapter 2 for further discussion). Unless otherwise specified, the time period for the analysis is assumed to be from 2013 to 2030. The key assumptions in the model and the specific steps by which the analysis proceeds are as follows:

STEP 1: ESTABLISHING KEY ASSUMPTIONS

- **PM2.5 target:** Reduce to 30 µg/m³ under the "structural adjustment scenario"
- **GDP growth:** Average of 6.8 percent growth (we assume that as economic growth potential changes, China's real GDP growth will gradually slow from roughly 8 percent in 2013–2014 to 5.5 percent in 2030)[17]
- **Average energy elasticity:** Defined as the ratio of the percentage change in energy consumption relative to a 1 percent increase in real GDP
  —Baseline scenario: 0.76, the same as it was between 2006 and 2012

—Structural adjustment scenario: 0.5 (That is, for every 1 percent growth in future GDP, the corresponding growth in energy consumption will be 0.5 percent, a rate approaching that of developed countries.)

- **Transportation elasticity:** Defined as the ratio of the percentage change in transport volume relative to 1 percent increase in real GDP
  - Baseline scenario: 0.9, the same as it was between 2008 and 2012
  —Structural adjustment scenario: 0.8 (That is, for every 1 percent growth in future GDP, the corresponding growth in the volume of transportation will be 0.8 percent, a trend that comports with the experience of other countries.)
- **Industry and construction growth:**
  —Baseline scenario: Secondary industry as a proportion of GDP remains constant.
  —Structural adjustment scenario: Secondary industry as a proportion of GDP will decline by 9 percentage points, which means that the noncoal emissions of secondary industries will also fall significantly.

STEP 2: DERIVING ENERGY CONSUMPTION AND TRANSPORT VOLUME GROWTH For both scenarios, we estimated annual energy consumption and transport volume growth based on the GDP growth estimate and assumptions for energy and transport elasticities.

STEP 3: ESTIMATING EFFECTS OF PM2.5 REDUCTION UNDER THE BASELINE SCENARIO We estimated that PM2.5 emissions per unit of coal consumed could be cut 70 percent by using clean coal technologies, including desulfurization and denitration. We also determined that a 78 percent reduction in vehicle emissions/km could be achieved by increasing the quality of petroleum products, fuel efficiency, and automotive emissions standards. Finally we found that the intensity of industrial and construction emissions per unit of output from noncoal sources could be reduced by 80 percent using technologies for soot control, clean production, and dust control.

Unfortunately, we also found that even if maximum efforts were expended to enhance emissions standards and deploy clean coal technologies, and the results from these efforts were fully realized, without additional structural adjustment, the 30 μg/m³ target cannot be met. That is, as long as the industry structure remains materially unchanged, energy consumption composition remains the same, and passenger vehicle ownership reaches 400 million units (according to expert forecasts), the PM2.5 level can be reduced only to 46 μg/m³.

STEP 4: CALCULATING CHANGES IN THE INDUSTRY, ENERGY, AND TRANSPORTATION STRUCTURES To meet the PM2.5 reduction target, the proportion of secondary industries must be substantially reduced as a proportion of GDP, conventional coal must be reduced and clean energy sources increased as proportions of total energy consumption, and the proportion of rail and subway transportation must be increased and road transport reduced. We considered various factors and combinations in our selection of pathways toward achieving these structural changes, including the availability of energy resources such as natural gas, hydropower, and wind power, as well as technical feasibility and international experiences—for example, in terms of rail transportation density and energy structure.

STEP 5: ESTIMATING FUTURE GROWTH IN ENERGY, AUTO-MOBILES, RAIL, AND SUBWAY TRANSPORT Based on the need to shift the total energy composition, we calculated growth rates for coal consumption and for various cleaner energy sources. We also calculated growth for rail, subway, and road transportation volumes, respectively, based on changes in the transportation structure. These growth rates served as the basis for calculating future changes in total rail and subway length and vehicle ownership. In the course of performing these calculations, we considered factors such as the decline in the automobile usage rate, improvements in rail transportation efficiency, and growth in the volume of other modes of transportation, such as air travel.

III. Results and Recommended Actions

Based on the model's simulations, we recommend specific actions that, taken together, can lead to the necessary emissions reductions to meet the PM2.5 target in 2030. Divided into two major categories, the first set of actions comprises technical environmental measures, most of which focus on end-of-pipe controls aimed at reducing the intensity of emissions from coal, transportation, industry, and construction. These actions include deploying clean coal technologies, improving the quality of petroleum products and fuel efficiency, increasing automobile emissions standards, and applying soot control, clean production, and construction dust control measures.

The second category centers on structural adjustments across the industry, energy, and transportation sectors. To reiterate, changes to the industry structure will enable the reduction of energy elasticity; increasing the proportion of clean energy will reduce emissions per unit of energy consumed; and decreasing road transportation will reduce emissions per unit of transportation volume.[18] Unless specified, these actions are all recommended for the 2013–2030 period.

1. Technical Environmental Measures

- Reduce emissions from conventional coal consumption by 70 percent using clean coal technologies, translating to an annual average reduction of 6 percent.
- Cut emissions per vehicle-km by 78 percent, which can be achieved by increasing fuel efficiency by 20 percent and cutting vehicle emissions per liter of fuel by 73 percent through more stringent fuel quality and automobile emissions standards, and through the wider deployment of electric and natural gas vehicles.
- Lower the intensity of noncoal emissions from industry and construction by 80 percent by controlling soot, adopting cleaner production methods, and controlling construction dust.

2. Structural Adjustment Measures

- Decrease secondary industry, including construction and manufacturing, as a proportion of GDP from 46 percent to 37 percent.
- Reduce the annual consumption of conventional coal by 14 percent and, more specifically, reduce the average annual growth in conventional coal consumption to 0.8 percent from 2014 to 2016, which means that conventional coal consumption should peak in 2016. (This would be nearly a decade earlier than the previous estimate of peak consumption around 2025.)

- Substantially reduce average annual growth in passenger vehicle ownership from the recent 20 percent to 6 percent and keep passenger vehicles to under 250 million, which is 150 million fewer than the current forecast of about 400 million vehicles.
- Increase total railway length by 60 percent and quadruple total subway length. More specifically, between 2020 and 2030, further increase rail and subway lengths by 60 percent and 230 percent, respectively.
- Substantially increase growth targets for clean energy—including natural gas, wind, nuclear, hydro, and solar—so that its proportion of total energy consumption rises from 14.6 percent to 37 percent between 2012 and 2020 and then to 46 percent in 2030. (This is much more aggressive than the official 20 percent nonfossil fuels energy target proposed for 2030.)

Assuming the outlined actions are fully realized, PM2.5 emissions reductions should be achieved in five areas, as shown in table 1.3.

The effects of structural adjustment on emissions reduction are not enumerated individually in table 1.3 (this is addressed in more detail in chapter 4). In the data below, the structural adjustment effects are already included in the percentages of emissions reductions from coal, transportation, and industry and construction. For example, the decline in the proportion of secondary industries will lead to a decline

TABLE 1.3
PM2.5 Reductions Based on Recommended Actions (2013–2030)

	Sources of PM2.5 (%)	% change	% contribution to reductions
Coal-related reduction	45	−74	62
Reduction in coal consumption		−22	18
Use of clean technologies		−70	44
Transportation-related reduction	22	−45	19
Construction and industry-related reduction	20	−47	17
Other pollution reduction	13	−6	1

Source: Author's PM2.5 model.

Note: Due to rounding, contribution to reductions does not add up to 100 percent.

in the energy elasticity. At a given GDP growth rate, energy and coal consumption growth can be reduced while simultaneously reducing growth in real industrial and construction output. As such, the decline in the proportion of secondary industry helps to cut emissions from coal, industry, and construction.

Furthermore, at a given rate of overall growth in energy consumption, changes to the energy structure will reduce conventional coal consumption, thereby also reducing emissions from coal burning. And finally, at a given rate of growth in overall transportation volume, changes to the transportation structure will reduce emissions originating from transportation.

Table 1.4 presents a more detailed picture of the changes in the various major indicators from 2012 to 2030, assuming the recommended actions are implemented from 2013 onward. Most important, if these recommended actions are fully realized, we anticipate that the average annual PM2.5 level will fall gradually from the current 65 μg/m³ to 30 μg/m³ by 2030 (see fig. 1.8).

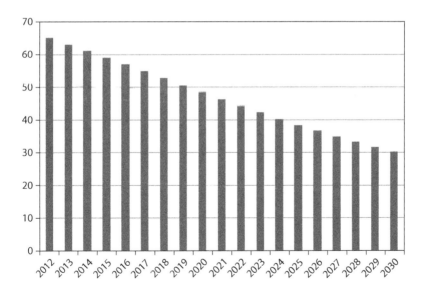

Figure 1.8 Annual PM2.5 reductions under recommended actions through 2030 (μg/m³). *Source:* PM2.5 control model.

TABLE 1.4

Estimates from PM2.5 Control Model: Emissions Reduction Road Map, 2012–2030

	2012	2013	2014	2015	2016	2017	2018	2019	2020	2021	2022	2023	2024	2025	2026	2027	2028	2029	2030
PM2.5	65.0	63.0	61.0	59.0	56.9	54.8	52.6	50.5	48.4	46.3	44.2	42.2	40.3	38.4	36.6	34.9	33.3	31.7	30.2
Coal-related emissions	29.3	27.8	26.3	24.8	23.3	21.7	20.2	18.8	17.4	16.0	14.8	13.6	12.5	11.5	10.5	9.7	8.9	8.2	7.5
Transportation-related emissions	14.3	13.9	13.6	13.2	12.9	12.5	12.2	11.8	11.4	11.1	10.7	10.3	10.0	9.6	9.2	8.9	8.5	8.2	7.8
Highway emissions	11.2	10.8	10.4	10.1	9.7	9.4	9.0	8.7	8.3	8.0	7.6	7.3	7.0	6.6	6.3	6.0	5.7	5.4	5.1
Railroad/subway emissions	1.4	1.4	1.4	1.5	1.5	1.5	1.5	1.5	1.5	1.4	1.4	1.4	1.4	1.4	1.4	1.4	1.3	1.3	1.3
Other transportation industry emissions	1.7	1.7	1.7	1.7	1.7	1.7	1.7	1.7	1.6	1.6	1.6	1.6	1.6	1.6	1.5	1.5	1.5	1.5	1.4
Construction and other industrial emissions	13.0	12.8	12.6	12.4	12.1	11.8	11.5	11.2	10.9	10.5	10.1	9.7	9.3	8.9	8.5	8.1	7.7	7.3	6.9
Other emissions	8.5	8.5	8.6	8.6	8.6	8.7	8.7	8.7	8.7	8.7	8.6	8.6	8.5	8.5	8.4	8.3	8.2	8.1	7.9
Coal emission intensity	36%	34%	31%	29%	28%	26%	24%	22%	21%	20%	18%	17%	16%	15%	14%	13%	12%	11%	11%
Auto emission intensity	100%	92%	85%	78%	72%	66%	61%	56%	51%	47%	43%	40%	37%	34%	31%	29%	26%	24%	22%
Industrial and construction emissions intensity	100%	92%	84%	77%	70%	64%	59%	54%	49%	45%	41%	38%	34%	32%	29%	26%	24%	22%	20%
Other emissions reduction measures	100%	93%	87%	81%	76%	71%	66%	62%	57%	54%	50%	47%	44%	41%	38%	35%	33%	31%	29%
Coal as a proportion of total primary energy consumption	67%	65%	63%	61%	59%	57%	55%	52%	50%	48%	45%	43%	41%	39%	37%	36%	34%	33%	32%
Total coal consumption (hundred million tons)	37	38	38	38	39	38	38	38	38	37	37	36	35	35	34	34	33	32	32
Road traffic as proportion of total transportation volume	55%	54%	54%	53%	53%	52%	51%	51%	50%	49%	49%	48%	47%	47%	46%	45%	45%	44%	43%
Road transportation growth		5.1%	5.2%	5.1%	4.9%	4.8%	4.6%	4.5%	4.3%	4.1%	4.0%	3.8%	3.7%	3.5%	3.4%	3.2%	3.1%	2.9%	2.8%

Vehicle utilization index	100%	99%	98%	97%	96%	95%	93%	92%	91%	90%	89%	88%	87%	86%	85%	84%	82%	81%	80%
Total number of vehicles (million)	110	117	124	132	140	149	157	166	176	185	195	205	215	226	236	247	258	269	281
Total number of passenger vehicles (million)	90	96	103	110	117	124	132	140	149	157	166	176	185	195	205	215	226	236	247
Passenger vehicle ownership growth		6.7%	6.9%	6.7%	6.6%	6.4%	6.3%	6.1%	6.0%	5.8%	5.7%	5.6%	5.4%	5.3%	5.1%	5.0%	4.9%	4.8%	4.6%
Passenger vehicle penetration†	6.7%	7.1%	7.5%	8.0%	8.4%	8.9%	9.5%	10.0%	10.6%	11.2%	11.8%	12.4%	13.1%	13.8%	14.5%	15.2%	16.0%	16.7%	17.5%
Population (million)	1,354	1,363	1,371	1,378	1,384	1,390	1,395	1,399	1,403	1,406	1,408	1,410	1,412	1,413	1,414	1,414	1,414	1,414	1,413
Railroad/subway transportation volume growth	9.2%	9.1%	9.5%	9.3%	9.1%	8.9%	8.7%	8.5%	8.4%	8.2%	8.0%	7.8%	7.6%	7.4%	7.3%	7.1%	6.9%	6.7%	6.5%
Railroad/subway transportation volume index (current level 100)	100	109	119	131	142	155	169	183	198	215	232	250	269	289	310	332	355	378	403
Railroad/subway transportation emissions index	100%	92%	85%	78%	72%	66%	61%	56%	51%	47%	43%	40%	37%	34%	31%	29%	26%	24%	22%
Construction/industry growth		7.5	7.7	7.4	7.0	6.7	6.5	6.2	5.9	5.6	5.3	5.1	4.8	4.6	4.3	4.1	3.9	3.6	3.4
Total energy consumption growth	3.9	4.0	4.2	4.0	3.9	3.9	3.8	3.8	3.7	3.5	3.4	3.2	3.1	3.0	2.9	2.8	2.6	2.4	2.1
Total transportation volume growth	6.2	6.2	6.4	6.3	6.2	6.0	5.9	5.8	5.7	5.5	5.4	5.3	5.2	5.0	4.9	4.8	4.7	4.5	4.4
GDP growth	7.8	7.7	8.0	7.8	7.7	7.5	7.4	7.2	7.1	6.9	6.8	6.6	6.4	6.3	6.1	6.0	5.8	5.7	5.5

Source: Estimates based on PM2.5 emissions control model.

Note: Penetration rate refers to per capita ownership—for example, a penetration rate of 10 percent means 100 vehicles/1,000 people.

CHAPTER TWO

Environmental Actions:
Necessary but Insufficient

AS THE MODEL'S results demonstrate, without comprehensive economic structural adjustments, China will not meet the national PM2.5 target by 2030 even if it achieves significant reductions in unit coal consumption, unit vehicle emissions/km, and unit noncoal emissions from industry and construction.

In this chapter, we present more detailed analysis of the environmental measures proposed and we show why relying solely on those actions, while necessary, will be insufficient to realize China's PM2.5 reduction objective.

I. Reducing Coal Consumption

Recall that an important component of the package of measures to control PM2.5 emissions is reducing emissions per unit of coal consumption by an average of 6 percent per year, or roughly a cumulative 70 percent by 2030. Such a drastic reduction may seem overly ambitious, but judging by past international experience, it appears feasible.

Consider Germany, for example. During the twenty-year period from 1991 to 2010, coal consumption in Germany rose 40 percent, while SO_2 and NOx emissions fell by 90 percent and 55 percent, respectively (see fig. 2.1). Average annual SO_2 emissions from thermal power plants fell from 2.4 million tons to under 100,000 tons. Desulfurization of thermal power plants played an important role in this rapid reduction.

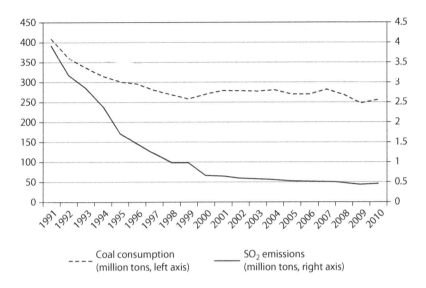

Figure 2.1 Historical reductions in coal consumption and SO₂ emissions in Germany. *Sources:* Author's estimates; OECD database; US Energy Information Administration.

In China, emissions reduction from coal consumption can be achieved by leveraging a combination of environmental measures and policies, specifically including (1) increasing the installation of desulfurization and denitration equipment; (2) substantially increasing the operating rate of such equipment; (3) shutting down small, inefficient thermal power plants; (4) promoting industry integration and upgrading; and (5) strengthening the design and enforcement of relevant regulations.

According to the China Electricity Council, 91.6 percent of thermal power plants had installed desulfurization equipment as of 2013. But to cut costs, some thermal power plants do not operate this equipment. While the MEP has officially reported a 95 percent operation rate of installed desulfurization equipment, experts believe the number may be lower.[1] Nevertheless, for all equipment installed, the MEP has calculated an overall desulfurization rate of 73.2 percent (operation rate multiplied by desulfurization efficiency).[2] Based on this data, only 64 percent (87 percent × 73 percent) of SO₂ emissions from thermal power have been controlled. This is also why SO₂ emissions remain relatively high, at 0.0063 tons/ton of coal consumption, nearly equivalent to the 1994 level in Germany and about four times Germany's 2010 level (see fig. 2.2).

China has not nearly achieved its desulfurization potential. We estimate that with more effort, up to 90 percent of SO_2 emissions can be controlled. To that end, we propose (1) raising the installation rate of desulfurization equipment in thermal power plants to 100 percent; (2) ensuring an operation rate of 95 percent or higher for installed equipment via stricter regulation and monitoring; and (3) remodeling flue gas bypasses to substantially increase overall desulfurization efficiency to 90 percent.

Denitration efforts are also very important. It is well known that burning coal also emits NOx emissions that undergo a chemical reaction to form secondary PM2.5 pollution. In the decade 2003–2012, NOx emissions increased 100 percent in China, with total volume now almost equal to that of SO_2 emissions.[3] Yet NOx emissions have not received similar levels of attention. As of early 2013, only 14 percent of China's thermal power plants had denitration facilities and equipment installed. To improve this situation, MEP authorities should mandate installation of denitration equipment within three years for

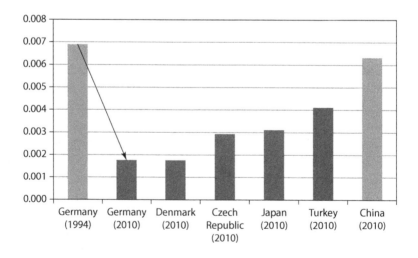

Figure 2.2 Comparison of SO_2 emissions (tons/ton of coal consumption). *Sources:* Author's estimates; OECD database.

all coal-burning power generators with an output of 200 megawatts (MW) or above. This is to push the actual denitration rate at major thermal power plants to 85 percent. The government should aim to have all thermal power projects install denitration equipment after 2015, which can be phased in over a few years.

Technologies such as bag-filtering dust precipitators should be promoted to reduce pollution from primary particulate matter (PM), which is made up of airborne particles such as dust, dirt, soot, smoke, and drops of liquid. This filtering technology enables one-time elimination of over 80 percent of primary PM pollution.

Coal used by thermal power plants accounts for just 52 percent of total coal consumed in China, with the rest going to industrial and residential uses, whose coal-related emissions also need to be strictly controlled. Although other sources of pollution besides coal-based thermal power are relatively dispersed and difficult to control, in our view, desulfurization and denitration facilities should, at a minimum, be required in key industries.

In addition, fees for SO_2 and NOx emissions should be raised substantially to force other industries and businesses to adopt emissions reduction measures. For example, the cement industry accounts for approximately 9 percent of total coal consumption nationwide each year and discharges 10 percent of the NOx and 5 percent of the SO_2. We recommend that cement production lines producing over 4,000 tons a day should meet an overall denitration target of 50 percent by the end of 2015. Flue gas denitration facilities used in the cement industry should also be promoted in other industries, while coal for residential use should largely be replaced by central heating systems fitted with emissions reduction equipment.

II. Reducing Auto Emissions

Even if the Chinese government enforces strong measures, such as automobile license plate auction systems and congestion fees to slow the growth in vehicle ownership, by 2030 the number of passenger vehicles is still very likely to reach 250 million, or 2.8 times the current level of about 90 million vehicles. Moreover, the total number of vehicles, including passenger vehicles, trucks, and public transportation vehicles, is estimated to reach 290 million by 2030.[4]

It is crucial for China to reduce unit vehicle emissions. We recommend reducing vehicle emissions/km by 78 percent through 2030, which will involve efforts in three areas.

First, implement stringent standards for petroleum products including gasoline and diesel fuel. Second, adopt stricter motor vehicle emissions standards—we estimate that when China's National V Fuel Standard (modeled after the Euro V standard) and National V Emissions Standard are fully implemented, the average vehicle emissions per unit of fuel consumed will fall 73 percent from current levels. Third, decrease fuel consumption per km by 20 percent by improving internal combustion engine technology and promoting wider adoption of electric and hybrid vehicles.

If these actions are taken, then by 2030—even as the total number of vehicles increases significantly—improvements in fuel quality, emissions standards, and fuel efficiency,[5] together with a 22 percent decline in the vehicle utilization rate (the distance driven by a single vehicle each year),[6] will result in 64 percent reduction in PM2.5 emissions from road transportation.

1. Impact of Fuel Standards Upgrade

In the process of fuel combustion, a portion of the sulfur in the fuel forms PM emissions directly, and another portion is first converted to SO_2 and then discharged. Part of this discharge is also converted into PM2.5 pollution in the atmosphere. Therefore, concurrent with efforts to control vehicle emissions, gasoline's sulfur content must also be reduced.

The National IV Fuel Standard stipulates that sulfur content must not exceed 50 parts per million (ppm), or 50 mg/kg, since sulfur content of 1 ppm means that 1 million kg of fuel contains 1 kg of sulfur. As of early 2013, the National IV standard had been implemented only in some urban areas, including Shanghai, Guangdong, Jiangsu, and Zhejiang, while most other areas were still on the National III standard, which stipulated that sulfur content cannot exceed 150 ppm. However, by the end of 2013, Beijing, Shanghai, and Jiangsu had all adopted the National V standard, which, in line with the Euro V standard, requires that sulfur content not exceed 10 ppm.

2. Impact of Auto Emissions Standards Upgrade

In addition to upgrading fuel quality standards, another important aspect of curbing transport-related emissions is to raise the vehicle emissions standard, which is measured by kg of emissions/km of driving distance. Because car replacement is largely voluntary, we estimate that it will take until 2022 for all vehicles on the road to meet the National V emissions standard. In fact, as of 2013, most vehicles on the road met only National III or IV standards.

Even so, gradually improving vehicle emissions standards will have an impact. Motor vehicles in China discharged a total of 46 million tons of pollutants in 2011, of which 35 million tons were carbon monoxide (which does not generate PM). Of pollutants associated with PM, direct PM emissions accounted for 5 percent, hydrocarbons accounted for 55 percent, and NOx accounted for 37 percent. According to the MEP's *Annual Report on the Prevention of Motor Vehicle Pollution*, gasoline-powered vehicles discharged 70 percent hydrocarbons, 30 percent NOx, and 10 percent direct PM, while diesel-powered vehicles discharged 90 percent direct PM, 70 percent NOx, and 30 percent hydrocarbons.

Based on the weightings above, the limits on light automobile pollutant emissions, and measurements in the National III, IV, and V standards, we estimate that upgrading from the National III to the National V standard will lead to an average 60 percent reduction in vehicle pollutant emissions (see table 2.1). These results were derived based on the weighted average of the various types of pollutants. In addition, 46 percent of the motor vehicles currently in use in China are on the National II Standard or below. This means that if these older vehicles are replaced, vehicle emissions will decline even further.

Once the National V emissions and fuel standards are completely phased in, by 2022, single-vehicle emissions will be 78 percent lower than the level in 2012. At the same time, if the government introduces incentives to encourage the purchase of low-emissions and energy-efficient vehicles, as well as electric vehicles (EVs), emissions will drop further.

3. Some Actions Taken, but More Can Be Done

On February 6, 2013, the General Administration of Quality Supervision, Inspection and Quarantine and the Standardization Administration of

TABLE 2.1
Vehicle Emissions Standards: National III to National V

		National III	National IV	National V	Emissions reduction National III to National V
Gasoline-powered vehicles	Carbon monoxide (g/km)	2.3	1.0	1.0	-57%
	Hydrocarbons (g/km)	0.2	0.1	0.1	-50%
	NOx (g/km)	0.15	0.08	0.06	-60%
Diesel-powered vehicles	Carbon monoxide (g/km)	0.64	0.50	0.50	-22%
	NOx (g/km)	0.50	0.25	0.18	-64%
	Direct PM (g/km)	0.50	0.025	0.0045	-99%

Sources: MEP; author's estimates.

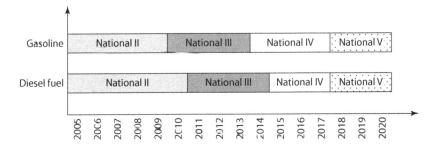

Figure 2.3 Implementation timetable for National V fuel standard. *Source:* Standardization Administration of China.

China issued a timetable for upgrading to the National IV and V standards for both diesel and gasoline fuel (see fig. 2.3). It essentially called for the National V gasoline and diesel standards to be rolled out by the end of 2013, with a phase-in period through the end of 2017. In our view, not only should such a timetable be strictly followed, but Beijing and other wealthy cities should lead the way with pilot projects to adopt more stringent standards. These cities should consider the Euro VI standard, for which the technology is apparently mature and has already been adopted in some European countries.[7]

Although upgrading to the National V standard should theoretically be completed in 2017, it will likely be 2022 before all vehicles are compliant with the new standard because it will take longer to replace vehicles that conform to the previous, lower standards. Nonetheless, this timetable is technically feasible. In major European countries, for example, it took four to five years for vehicles to upgrade from Euro III to Euro IV (equivalent to China's National III and IV standards), and another three to four years to adopt the Euro V standard.

III. Reducing Industry and Construction Emissions

Noncoal emissions from industry and construction are relatively dispersed in the atmosphere. Generally acknowledged sources include industrial soot (dust), VOCs, and construction dust. We argue that the intensity of emissions for these three types of pollutants can be reduced by about 80 percent with existing technologies. This will also lead to an 80 percent reduction of unit noncoal emissions from industry and construction by 2030.

1. Industrial Soot

According to MEP's *2011 Annual Environmental Statistics Report*, soot (dust) represented the third-largest air pollutant after NOx and SO_2, with a total volume of 12.8 million tons, about 86 percent of which were industrial emissions. The nonmetal mining industry (primarily cement) accounted for 27 percent of soot emissions, electricity and heating (primarily thermal power) accounted for 21 percent, and the ferrous metal smelting and pressing industry (primarily steel) accounted for 20 percent.[8]

Pollutants from the thermal power and steel industries are primarily due to coal combustion and peripheral operations, but the cement industry is different. Of the nearly 3 million tons of dust (soot) discharged by the cement industry annually, the majority is not generated from burning coal. Rather, soot is generated in a series of processes extending from cement kiln to drying, cooling, crushing, milling, storage, and transportation. Particularly in the raw mill and clinker mill production segments, dust generation per unit of product can be as high as 1,000 kg/ton, far higher than the 75–120 kg/ton in the combustion process of cement kilns using the latest drying method.[9] If efforts are made to address these production segments, industrial soot levels can be substantially reduced.

The previous PM emissions limits on the cement industry, set at 30–50 mg/m³ for each production segment, were initially put in place in 2005. But the new standards, issued in 2013, call for an average of at least 40 percent reduction of various emissions of soot from China's cement industry (see table 2.2). In addition, for cement companies in key regions across China, including 47 cities where environmental capacity is weak,[10] the new regulations set stricter standards. Assuming that the new standard is fully implemented nationwide, cement-related emissions should decline by nearly 60 percent from the 2012 level.

Our study concluded that there is room to set higher standards after 2015. Using the current soot removal technology, PM emissions in cement kilns can already be limited to under 20 mg/m³ using

TABLE 2.2
Comparison of PM Emissions Standards for Cement Industry

Production process	Production equipment	Particulate matter (mg/m³)		
		GB 4915-2004 Implemented 01/01/2005	GB 4915-2013 General standard, implementation for 2013	GB 4915-2013 Standard for businesses in key regions, implementation for 2013
Mining	Crushers and other ventilation and production equipment	30	20	10
Cement making	Cement kilns and in-line kiln/raw mills	50	30	20
	Dryers, drying mills, packaging machinery and ventilation, and production equipment	30	20	10
Bulk cement transfer stations and production of cement products	Cement storage silos and other ventilation and production equipment	30	20	10

Source: MEP.

high-performance static electricity or cement kiln dust collection equipment. Going forward, if cement production uses state-of-the-art technology—such as the new membrane filter media dust collectors—and adheres to the most stringent environmental standards, it is possible for soot and dust PM concentrations to be kept below 10 mg/m³. Therefore, the 80 percent emissions reduction of soot is achievable.

2. *Industrial VOCs*

VOCs are a significant contributor to both PM and smog, but they differ from SO_2 and NOx emissions in that VOC emissions are dispersed and are broadly present in the petrochemical, automotive spray-painting, and printing industries. When exposed to sunlight, the combination of VOCs and NOx generates two types of pollutants: (1) secondary PM, a component of PM2.5; and (2) ozone, one of the major sources of photochemical smog. Industrial VOCs, in particular, constitute 75 percent of VOC emissions from immobile sources.[11]

However, industrial VOCs happen to be a pollutant that can be effectively controlled through regulation and enforcement. Although China so far has not regulated industrial VOC emissions, the MEP is currently creating industry-specific VOC standards. This presents an opportunity to ensure that the standards substantially cut VOC emissions. In particular, over 90 percent of industrial VOC emissions come from a few industries—coatings, solvents, fossil fuel processing, and transportation—which are generally organized emissions. That means they are highly concentrated and are easily collected and processed. Moreover, technologies to curb these emissions are mature and available.

Consider paints and coatings, for example, which account for 58 percent of industrial VOCs. China has apparently already mastered the production technology for waterborne anticorrosive coatings, which can cut the use of organic solvents by 60–80 percent, thereby reducing associated VOC emissions, according to the Compilation Notes of the MEP's *Technical Policies to Prevent Volatile Organic Compound Pollution*. This technology can be applied to the coating requirements for numerous products, including bridges, automobiles, and shipping containers.

In 2011, the Ministry of Industry and Information Technology began promoting the coating technologies described above, but the targets for 2013 were set very low, calling for penetration rates below 15 percent. If these technologies reach near full penetration by 2030, VOC emissions can be reduced by 70 percent.

Finally, the petrochemical industry accounts for nearly 20 percent of industrial VOC emissions. The results of several pilot projects show that use of leak detection and repair technologies can decrease petrochemical VOC emissions by 90 percent. In the storage link of the production process, floating roof tanks can reduce emissions by 85–96 percent compared to fixed roof tanks, but fixed roof tanks are still being widely used for storing diesel fuel. If major urban areas simply adopted existing technologies, VOC emissions could be significantly curtailed. In addition, if the Shanghai pilot project on floating roof tanks were expanded nationally, VOC emissions from China's petrochemical industry could be reduced by 80 percent.

3. Construction Dust

Construction dust is also a source of pollutants. Although it is difficult to provide precise measurements, based on MEP's *Notice Concerning the Effective Control of Urban Dust Pollution* and historical experience with measures adopted in Beijing and Shanghai, a slew of actions can be taken to eliminate such dust. These include sealing vehicles transporting dust-prone materials, using sealed storage in storage yards, reducing exposed work areas, prohibiting cement mixing on site, expanding urban greenery, and sweeping and cleaning roads, among other actions. Shanghai also has sealing requirements for vehicles transporting waste soil, as well as GPS systems and online monitoring of construction sites throughout the city. If effective monitoring and control measures can be rolled out nationally, construction dust's contribution to PM2.5 can be markedly reduced.

IV. Assessing Overall Results from Environmental Measures

While the objective is to reduce the national PM2.5 level from 65 $\mu g/m^3$ to 30 $\mu g/m^3$ by 2030, we recognize that this task is much more difficult than simply cutting emissions by 35 $\mu g/m^3$. The main reason is because the Chinese economy—GDP calculated at constant prices—is estimated to grow 227 percent between 2013 and 2030, assuming average growth of 6.8 percent. Given this assumption, if the energy and transportation elasticity coefficients, the intensity of emissions per unit of energy consumed, and per unit of transportation volume remain unchanged, PM2.5 emissions will rise 166 percent, to 173 $\mu g/m^3$.

TABLE 2.3

Simulated PM2.5 Levels under Baseline Scenario

	PM2.5 Reduction Results (μg/m³)
"Do nothing" scenario (assuming all policies and economic structure remain unchanged)	173
Environmental measures' impact (baseline scenario):	
70 percent reduction in emissions per unit of coal consumed	120
78 percent reduction in vehicle emissions/km	100
75 percent reduction in the intensity of emissions from secondary industries and other sources	46

Source: Author's PM2.5 control model.

Note: The model assumes that the proposed measures are implemented according to the sequence in the table above. In reality, however, all measures are implemented simultaneously.

Our model demonstrates that under the baseline scenario, the average urban PM2.5 level will fall to only 46 μg/m³ (see table 2.3). Therefore, under the assumption that economic growth is unlikely to decline precipitously and that end-of-pipe measures are fully realized, China will then need to resort to structural adjustments to reach the finish line. It is to this topic that we turn in the next chapter.

CHAPTER THREE

Structural Adjustment: The *What* and the *How*

AS IS CLEAR from our analysis so far, China needs to substantially alter its industry, energy, and transportation structures to meet the PM2.5 reduction target (see figure 3.1). Based on our model's simulations, economic structural adjustments between 2013 and 2030 should aim to decrease secondary industry as a proportion of GDP by 9 percentage points (from 46 percent to 37 percent), increase the proportion of clean energy by 32 percentage points (from 14.6 percent to 46 percent), and raise urban rail transportation as a proportion of total travel by 23 percentage points (from 10 percent to 33 percent). Only when these structural adjustments are combined with environmental measures will it be possible to meet the 2030 target.

The first part of this chapter discusses how these structural adjustments will affect PM2.5 levels. The subsequent three sections offer detailed analysis of the actions, policies, and technical feasibility of the required structural adjustments. The last section presents quantitative estimates of the extent to which each type of adjustment contributes to reducing PM2.5 emissions.

I. Types of Structural Adjustments Needed

1. Industry

Broadly speaking, changing industry's structure will affect energy consumption elasticity, transportation elasticity, and industrial and construction output, thereby affecting PM2.5 levels. Specifically, if the

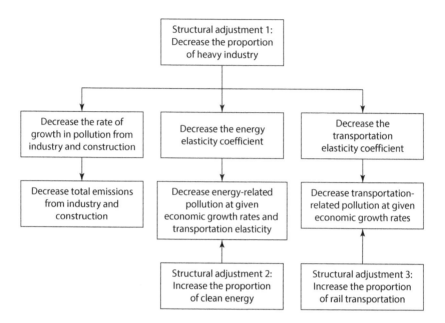

Figure 3.1 Structural adjustments' impact on PM2.5 emissions. *Source:* Author's PM2.5 emissions control model.

proportion of secondary industry in an economy is relatively high, the energy elasticity in that economy is generally also relatively high—meaning there is a positive correlation. This is because energy consumption per unit of output in manufacturing and heavy industry is far higher than in other industries—about four times higher than in the services sector. Consequently, China's energy elasticity in the decade 2003–2012, which coincided with rapid industrialization, averaged around 0.8. Meanwhile, OECD countries' energy elasticity stood around 0.5 over the same period.

Put another way, at a given GDP growth rate, if the proportion of secondary industry is relatively high, energy consumption growth also tends to be relatively high. All other conditions being equal, higher energy consumption growth naturally leads to faster growth in pollution emissions. Conversely, decreasing the proportion of secondary industry in the economy can help cut energy consumption elasticity, which in turn reduces energy consumption growth and associated pollution emissions.

In addition, the proportion of secondary industry in an economy is also positively correlated with transportation elasticity. This is because the intensity of transportation is far higher in secondary industries than in other industries. The United States, for example, saw its real GDP expand 167 percent from 1970 to 2000, while its total cargo transport volume grew by only 73 percent because its services sector as a proportion of the economy expanded substantially. As such, China's transportation elasticity coefficient averaged around 0.9 over 2008 to 2012, while it was about 0.5 for OECD countries in the decade 2002–2013.

So as with energy elasticity, for a given rate of GDP growth, if the proportion of secondary industry is relatively high, then so will be transportation volume growth. All other conditions being equal, higher transportation volume growth results in faster growth in air pollution emissions. Conversely, if the proportion of secondary industry declines, that should lower transportation elasticity, thereby slowing the growth of transportation-related emissions.

Finally, the proportion of secondary industry in GDP will also affect noncoal emissions from manufacturing and construction, including industrial soot, VOCs, and construction dust, which are the third-largest source of PM emissions behind coal combustion and transportation-related emissions. For a given rate of GDP growth, if the other conditions remain the same, higher growth of secondary industry will lead to faster growth in emissions from manufacturing and construction. Conversely, a decline in the proportion of secondary industry can help lower growth in these types of noncoal emissions.

2. Energy

Changes in the energy structure will affect PM2.5 emissions per unit of energy consumed, or emissions intensity, usually measured in tons of coal equivalent (tce). Because the emissions intensity of clean energy sources is far lower than that of conventional coal, reducing conventional coal as a proportion of total energy consumption will bring down PM2.5 emissions. For example, the emissions intensity of SO_2 from natural gas is just one-twentieth to one-tenth that of conventional coal.

If coal as a proportion of energy consumption is reduced by 30 percentage points, and clean energy is increased by an equivalent proportion, energy-related emissions from total energy consumption would fall by approximately 25 percent.

3. Transportation

Altering the transportation structure will affect PM2.5 emissions per unit of transportation volume. Increasing the ratio of rail transport relative to total transportation volume, particularly subways as a proportion of total urban transportation, supports PM2.5 reductions. That is because the emissions intensity of rail transportation—the energy consumed and the PM2.5 emissions generated per ton-km or passenger-km—is far lower than that of road transportation. For example, the unit emissions per passenger-km of rail transport are one-tenth that of private automobiles. If subway transportation as a proportion of total transportation increases by 30 percentage points and the proportion of private automobile travel is reduced by the same level, passenger travel-related emissions will decrease by about 27 percent in terms of given traffic volumes.

In the next several sections of this chapter, we examine the magnitude of each type of structural adjustment as part of our recommended actions. Estimates of the effectiveness and feasibility of these adjustments in reducing emissions also follow.

II. How Much Industry Structural Adjustment Is Needed?

Based on our model simulations, total secondary industry as a proportion of GDP should decline from 46 percent to 37 percent, or by 9 percentage points, which in turn means that the average annual growth in secondary industry should slow to about 80 percent of the GDP growth rate. If the 9 percentage point reduction is achieved, compared to the scenario in which secondary industry as a proportion of GDP remains unchanged, PM2.5 emissions can be reduced by about 6 $\mu g/m^3$ by 2030. Details of the various elasticities used in our simulation follow.

First, in terms of energy elasticity, our analysis shows that if the proportion of secondary industry falls an average of 9 percentage points through 2030, then energy elasticity will decrease by an average of about 0.15 percentage points, from 0.65 to approximately 0.5, and transportation elasticity will decrease an average of 0.1 percentage points, from 0.9 to about 0.8. In terms of growth of the secondary industry itself, assuming the 6.8 percent average GDP growth between 2013 and 2030, if the proportion of secondary industry declines by a cumulative 9 percentage points, then the sector will see an average growth of 5.6 percent, 1.2 percentage points lower than GDP growth. This means the annual rate of noncoal emissions reduction from secondary industries (construction and heavy industry) can be accelerated by 1 percentage point.

Some may question whether the proposed reduction in secondary industry is overly ambitious. Indeed, it is difficult to imagine such drastic decline in industry after a decade in which China experienced an *increase* in the proportion of secondary industry. However, historical international experiences give cause for optimism (see table 3.1). These cases demonstrate that after reaching $8,000 per capita GDP (calculated using constant 2005 prices based on purchasing power parity), these economies saw a period of sustained and relatively rapid structural adjustment, resulting in the decline of the proportion of secondary industry by an average of roughly 10 percentage points.[1] Their experiences suggest that it is possible for China to achieve the 9 percentage point reduction over roughly an eighteen-year period.

TABLE 3.1
Declines in Secondary Industry in OECD Countries

OECD Country	Period	Change in proportion of secondary industry (percentage points)
United Kingdom	1985–2003	−14.3
Russia	1991–2009	−14.0
Japan	1991–2009	−11.4
Germany	1978–1996	−10.7
Spain	1975–1993	−10.3
United States	1981–1999	−10.1
Italy	1974–1992	−9.8
France	1976–1994	−8.8
Canada	1974–1992	−8.4
South Korea	1991–2009	−5.8

Sources: World Bank; author's estimates.

Of course, simply relying on international experience is insufficient. There are certain factors that can drive a decline in the proportion of secondary industry, chief among them changes in demographics, policies, and demand elasticity.

Both theoretical and empirical studies demonstrate[2] that an aging population increases a country's consumption rate and lowers its investment rate, and a decline in the investment rate is most likely associated with a decline in the proportion of secondary industry. In addition, policies that change the relative cost of investing in secondary and tertiary industries, such as adjusting taxes and land prices, can cause a shift in resources from secondary to tertiary industries. Finally, as the number of physical products owned per capita increases (i.e., the penetration of goods increases), consumer demand elasticity for physical products decreases as demand elasticity for services rises.

The effects of the first two factors—demographics and policies—can be simulated directly using a CGE model. But the effects of the third factor cannot be simulated directly in a CGE model due to limitations from fixed elasticity coefficients, though the historical experiences of developed countries can shed some light on this factor.

Using the CGE model, we estimate that the demographic factor— that is, a decline in the working-age population and an increase in the older population—will lead to a reduction of secondary industry as a proportion of GDP by about 2 percentage points. The remaining reductions must be achieved through additional policy initiatives and demand elasticity changes, such as:

- Fully converting the business tax to a value-added tax (VAT) in the services sector, keeping the overall tax burden the same. This essentially means a *de facto* tax increase on secondary industries, with the resulting fiscal revenue used to offset taxes on tertiary industries.
- Increasing resource tax rates to slow the growth of heavy industry.
- Imposing higher fees on pollution emissions to incentivize reallocation of resources into the services sector.
- Reforming the supply and pricing mechanisms for designated industrial land and substantially raising the cost of that land. (At present, the average cost of industrial land in China is only 13 percent the cost of commercial land and 16 percent the cost of residential land. In contrast, these ratios range from one-third to 50 percent in most developed countries.)

If the policies described above are sufficiently aggressive and fully enforced, and if they are combined with changes in the demographic structure, the proportion of secondary industry may be reduced by 3 percentage points, according to our estimates (see chapter 8 for further discussion). The remaining 6 percentage points can be achieved via changes in demand elasticity as income growth generally leads to more consumption of services, according to the empirical experience of developed economies. The bottom line: our model suggests that it will be feasible to significantly curtail the proportion of secondary industry within the given time frame.

III. Energy Structure: Less Coal, More Everything Else

Of the various undertakings we have proposed, the most challenging will be curbing the consumption of conventional coal, which, for our purposes, does not include coal used for generating electricity via coal gasification. Nonetheless, the key objective is to achieve peak coal consumption in 2016 and then see a reduction of consumption by 2017, with a cumulative decline of 18 percent from 2017 to 2030. This translates into an average annual decrease of 0.8 percent, or a total of 14 percent from 2013 to 2030.

Achieving peak coal consumption by 2016 may seem quite ambitious, given that as of 2013, many coal industry experts were still projecting 4–6 percent growth in coal consumption for the following several years. Even conservative forecasts estimate that conventional coal will continue to grow 2–3 percent annually through 2020. Numerous studies, including those from Chinese government institutions, projected that coal consumption would peak between 2020 and 2030 (see table 3.2).

Although it will be very difficult to reach peak coal consumption five to ten years earlier than these forecasts, it is our view that as long as the Chinese government is genuinely determined to improve the environment and vigorously executes economic reforms, such as substantially increasing the resource tax and fees for coal emissions, meeting our target is feasible.

Of course, achieving an average annual decline of 0.8 percent does not mean that coal consumption will drop immediately. In fact, our model uses a different growth rate for each year from 2013 to 2030.

TABLE 3.2
Forecasts of Peak Coal Consumption

Forecast Source	Peak year
Author's current forecast	2016
Author's previous forecast	2020
International Energy Agency	2020
Hao Pengmei	2020
Alibaba	After 2020
Hu Angang et al.	2025
Lawrence Berkeley National Laboratory	Late 2020s
Du Xiangwan	2030
Brad Plumer	2030

Sources: International Energy Agency, *World Energy Outlook*, 2012; Hao Pengmei, "Forecast of Medium- and Long-Term Development Trends in China's Coal Industry," *China Coal*, 38, 2012; Alibaba, http://info.china.alibaba.com/detail/1067462601. html; Hu Angang, "China 2030: The March Toward Collective Prosperity," *Renmin University of China Press*; Du Xiangwan, speech at the 2009 High-Level Forum of Chinese Energy Companies; Brad Plumer, "China Now Burning as Much Coal as the Rest of the World Combined," *Washington Post*, January 2013; Lawrence Berkeley National Laboratory, "China's Energy and Carbon Emissions Outlook to 2050," April 2011.

Notes: Author's current forecast was made when the research project launched in early 2013; previous forecast was based on Deutsche Bank estimates made in 2012.

Assuming that clean coal technologies such as desulfurization and denitration are fully applied, and central heating is aggressively promoted, slow growth in coal consumption will persist until 2016, but a substantial decrease should start in 2017. This forecast also takes into account the very low base of clean energy supply, which is unlikely to replace conventional coal in a meaningful way in the near term (see table 3.3).

In addition to controlling coal consumption and deploying clean coal technologies, another important component of our proposed actions is accelerating the penetration of clean energy, which, for this purpose, is defined as including conventional and unconventional natural gas, wind, nuclear, hydro, and solar power (see tables 3.4 and 3.5).

Compared to our previous forecast, which was based on 2012 estimates from Deutsche Bank's research team, our current forecast under the structural adjustment scenario adjusts the energy structure more substantially. Based on the latest forecast, conventional coal

TABLE 3.3
Estimates of Conventional Coal Consumption Growth (2012-2030)

	2012	2013	2014	2015	2016	2017	2018	2019	2020	CAGR, '13-'20
Previous forecast*	1.2%	4.0%	3.5%	3.0%	2.5%	1.5%	1.0%	0.5%	0.0%	2.0%
Current forecast†	1.2%	1.5%	1.2%	0.9%	0.4%	-0.1%	-0.4%	-0.7%	-1.0%	0.2%

	2021	2022	2023	2024	2025	2026	2027	2028	2029	2030	CAGR, '21-'30
Previous forecast	-0.5%	-1.0%	-1.0%	-1.0%	-1.0%	-1.0%	-1.0%	-1.0%	-1.0%	-1.0%	-1.0%
Current forecast	-1.3%	-1.7%	-1.7%	-1.7%	-1.7%	-1.7%	-1.7%	-1.7%	-1.7%	-1.7%	-1.7%

*Previous forecast was based on Deutsche Bank estimate in 2012, which assumed limited structural adjustment to energy composition.

†Current forecast refers to the structural adjustment scenario in which aggressive economic restructuring is undertaken.

Source: Author's estimates.

TABLE 3.4
Forecast of Energy Mix Composition Through 2020

	Status in 2012	Current forecast	Previous forecast	Forecast adjustment (percentage points)	CAGR, '13–'20 (previous)	CAGR, '13–'20 (current)
Coal	66.6%	49.9%	57.4%	−7.5	2.0%	0.2%
Oil	18.8%	22.8%	23.5%	−0.7	6.9%	6.4%
Clean energy	14.6%	27.3%	19.1%	8.2	7.5%	12.2%
Wind	0.9%	2.5%	1.2%	1.3	7.9%	18.3%
Solar	0.04%	0.4%	0.2%	0.2	27.1%	38.6%
Natural gas	5.2%	12.0%	8.5%	3.5	10.5%	15.4%
Nuclear	0.9%	3.2%	2.0%	1.2	15.5%	22.4%
Hydro	7.6%	9.2%	7.2%	2.0	3.2%	6.4%
Total*	100%	100%	100%			

* Figures are rounded; totals may exceed 100% Clean energy refers to the sum of wind, solar, natural gas, nuclear and hydro.

TABLE 3.5
Forecast of Energy Mix Composition Through 2030

	Status in 2013	Current forecast	Previous forecast	Forecast adjustment (percentage points)	CAGR, '13–'30 (previous)	CAGR, '13–'30 (current)
Coal	66.6%	31.7%	39.2%	-7.5	0.3%	-0.8%
Oil	18.8%	22.6%	24.2%	-1.6	4.8%	4.4%
Clean energy	14.6%	45.7%	36.6%	9.1	8.8%	10.1%
Wind energy	0.9%	5.1%	3.6%	1.5	11.7%	13.9%
Solar energy	0.04%	3.1%	1.8%	1.3	27.7%	31.6%
Natural gas	5.2%	18.0%	14.2%	3.8	9.3%	10.7%
Nuclear energy	0.9%	8.0%	7.0%	1.0	16.1%	17.0%
Hydropower	7.6%	11.5%	10.0%	1.5	4.9%	5.7%
Total	100%	100%	100%			

Note: Forecast based on annual average growth of 3.4 percent in total energy consumption and average annual decrease of 0.8 percent in coal consumption.

Source: Author's estimates.

consumption as a proportion of total energy consumed will decline from 67 percent in 2012 to 32 percent in 2030 (it was 42 percent in the previous forecast). At the same time, the proportion of clean energy will increase from 14.6 percent in 2012 to 27 percent in 2020 and then to 46 percent in 2030. The average annual growth of clean energy from 2013 to 2030 should reach 12 percent, 4 percentage points higher than the original forecast.

For natural gas, we revised our estimated average annual growth for 2013 to 2020 from 10.5 percent to 15.4 percent. This is because, in our view, natural gas could well account for 18 percent of total energy consumption by 2030. Although solar accounted for less than 0.1 percent of total energy consumption in 2012, it is projected to expand at an average rate of 32 percent, accounting for 3 percent of total energy consumption by 2030. We also raised our estimated average annual growth for wind and nuclear energy to 14 percent and 17 percent, respectively, from 12 percent and 16 percent in the previous forecast. Considering the constraints on resource availability and environmental impact, growth of hydropower was adjusted downward to an average growth of 5.7 percent.

While these forecasts appear ambitious, a quick comparison with international experience may help to put these objectives in context. For instance, in 2012, clean energy accounted for only 14.6 percent of total energy consumption in China, while the OECD average had already reached 42 percent. In France, clean energy accounted for as much as 62 percent of total energy consumption (see figure 3.2). The differences across economies lie not only in the availability of natural resources in each country, but also reflect the degree of policy support and the amount of investment in clean energy. Based on the European examples, as well as factors such as the maturation of coal gasification technology, declining cost of solar, and the availability of unconventional natural gas resources such as shale gas and coal-bed methane (CBM), it is possible, in our view, to boost clean energy in the total energy mix to 27 percent by 2020 and to 46 percent by 2030.

The feasibility of adjusting the energy composition accordingly will be discussed in the following sections. But first, we briefly examine the various clean energy sources that will be important drivers in shifting China's overall energy composition.

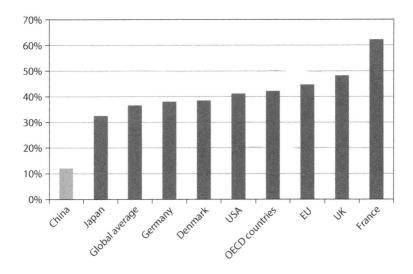

Figure 3.2 Clean energy as a proportion of primary energy consumption in OECD countries. *Sources:* Author's estimates; *BP Statistical Review of World Energy* (2012).

1. Natural Gas

Natural gas has the greatest potential to replace coal, and switching from coal to natural gas can bring enormous environmental improvements. Official projections from the National Development and Reform Commission (NDRC) show that natural gas consumption was projected to rise from 130 billion cubic meters (bcm) in 2011 to 230 bcm by 2015 as part of the twelfth FYP for Natural Gas Development (2011–2015), an average of 15 percent annual growth. We estimate that natural gas will continue to grow rapidly, accounting for 12 percent of total primary energy consumption by 2020 and 18 percent by 2030.

Assuming all other factors remain constant, if that amount of natural gas offsets what would have been a corresponding increase in coal consumption of an equivalent calorific value, then it would lead to annual CO_2 reductions of 520 million tons and SO_2 reductions of 5.8 million tons, equal to 28 percent of national SO_2 emissions in 2011. In terms of transportation, a Tsinghua University study suggests that natural gas–powered public transportation vehicles emit just 0.005 μg/km of

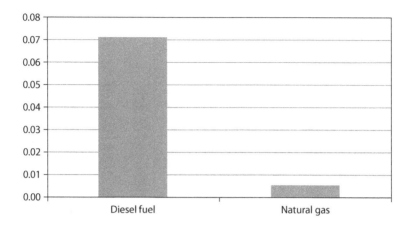

Figure 3.3 PM2.5 emissions from public transportation vehicles using different fuels (g/km). *Source:* Tsinghua University Department of Environmental Science and Engineering and SEPA, "Study on Planning for the Control of Motor Vehicle Pollution Emissions in China," 2001.

PM2.5, about 93 percent less than vehicles powered by National IV Standard diesel fuel (figure 3.3).[3]

We believe the targets above are realistic, for several reasons. For one, the Chinese government and businesses have invested extensively in natural gas R&D, exploration, production, transportation, and infrastructure. Beijing pledged that by the end of 2015, major gas supply arteries comprised of the West-East pipeline, the Sichuan-Eastern China pipeline, and the Shaanxi-Beijing pipeline will be complete, supplying the needs of about 18 percent of China's urban population, or about 250 million people. In addition, more natural gas filling stations will be built and more subsidies will be introduced to encourage increasing gasification of public bus and taxi fleets.

Two, China has abundant natural gas reserves, including recently discovered vast shale gas reserves, as well as long-standing CBM resources. Based on a domestic oil and gas resource assessment in 2010, China estimated its technically recoverable natural gas resources at approximately 3.2 trillion cubic meters (tcm), about 1.08 tcm for CBM at a depth of 2,000 meters or less, and 2.5 tcm for shale gas. Another estimate, from the US Department of Energy and the International

Energy Agency in 2013, ranks China's shale gas reserves as the world's largest, surpassing those of the United States.

Three, although shale gas and CBM development remains in the embryonic stage and current costs are high relative to those in the United States, their potential is evident from the extraction that has already occurred. China has drilled over 100 shale gas wells in Sichuan, Chongqing, and other regions. According to data from Hongyu Information, shale gas production in China reached 1.3 bcm in 2014 and will likely hit the government's production target of 6.5 bcm in 2015.[4] At the corporate level, China's largest oil company, China National Petroleum Corporation, has already announced plans to produce 1.5 bcm of shale gas by 2015, 20 bcm by 2020, and 50 bcm by 2050. In addition, cooperation with foreign companies will help accelerate the cost reduction of extracting unconventional gas. For example, Shell China has stated that it aims to cut operating expenses for the Fushun-Yongchuan shale block by 40–60 percent, which should theoretically slash the cost of drilling a single horizontal well from about $12 million to $4 million.

2. *Nuclear*

Nuclear power supplies 15 percent of the world's electricity, while in China it supplies less than 2 percent. International and domestic experiences demonstrate that nuclear power development depends significantly on state and policy support. In its twelfth FYP for energy development, the Chinese government proposed tripling the installed capacity of nuclear power from roughly 11 gigawatts (GW) in 2010 to 40 GW within five years. Experts at the China Energy Research Society have also suggested that China could reach 80 GW of installed capacity by 2020. However, since the Fukushima accident in Japan, China has slowed the pace of building its nuclear power fleet.

Still, this industry has significant growth potential for several reasons. First, China's enormous future energy demand urgently requires the development of renewable energy. But because solar and wind power capacity still cannot be reliable baseload power substitutes to replace coal, nuclear can serve that role. Second, China has now essentially mastered the more sophisticated and safer third-generation nuclear power technology, based on the AP1000 reactor technology transferred from the US company Westinghouse, which should further alleviate safety concerns.

Taking into account the current cautious policy environment, we estimate that nuclear energy will nonetheless expand at a relatively healthy clip and could account for 3.2 percent of energy consumption by 2020. That will amount to slightly less than 80 GW of installed capacity.

3. Hydropower

China's hydropower resources total 694 GW. We expect hydropower capacity will meet the twelfth FYP's target for renewable energy in 2015 and will continue to grow through 2020. This would mean doubling the current 220 GW of installed capacity to 420 GW and boosting hydropower in the total energy mix from 7.6 percent to 9.2 percent. However, due to resource limitations such as water availability and various environmental impacts, hydropower's growth will likely slow after 2020.

4. Wind

Based on the China Meteorological Administration's fourth wind energy survey, China's wind power potential exceeds 2 billion kilowatts, or 2 terawatts. Therefore, it is entirely possible for China to achieve the installed capacity target of 400 GW by 2030 through multipronged efforts such as strengthening the power grid infrastructure, better planning of resource utilization, optimizing power transmission, and reforming electricity pricing mechanisms.

Moreover, although still in its infancy, offshore wind power holds significant untapped potential. Offshore wind technology for water depths of 5–25 meters is already relatively mature, though technological improvement is needed for deeper waters. In fact, the NDRC's Energy Research Institute (ERI) notes that 500 GW of offshore wind resources can be tapped for development using existing technologies (between 5 and 25 meters).[5] By 2020, the installed capacity of grid-connected wind power is projected to reach 200 GW, up from 45 GW in 2011, generating more than 390 billion kilowatt-hours (kWh) of electricity a year.

This 2020 target, if achieved, means wind power should account for over 2 percent of total energy consumption, which aligns with our current forecast of 2.5 percent. For 2030, ERI has projected 400 GW of installed capacity, which comports with the 14 percent average annual growth proposed in our model.

5. *Solar*

In its twelfth FYP for renewable energy, the Chinese government pro-posed a target of 50 GW of installed solar power by 2020, which rep-resents an enormous increase from the 860,000 kW available in 2010. This means that from 2010 to 2030, solar energy must average more than 50 percent growth per annum. Taking into account that solar power still needs considerable state subsidies, however, we forecast lower average growth, of 39 percent, from 2013 to 2030.[6]

In sum, the short-term (2015) targets for clean and renewable energy in our current forecast are consistent with both the sectoral targets the central government has announced and with the forecasts of industry experts. Our medium- and long-term targets are also achievable, both from a technical perspective and in terms of resource potential. Ultimately, whether our relatively aggressive targets can be achieved are dependent on the government's ability to introduce strong and effective policies and incentives, as well as the availability of financing—topics to which we'll turn later.

IV. Curtailing Automobile Ownership

Lowering transportation-related PM2.5 emissions by 50 percent, as proposed in our model, will require focusing on several key areas. First, curtail automobile ownership to reduce growth of road trans-portation; second, cut single-vehicle emissions by 73 percent by raising fuel quality and emissions standards; third, increase fuel efficiency by 20 percent; and fourth, maintain a 7.1 percent average annual growth in rail and subway transportation.

Based on the parameters for each of the above factors, we esti-mate that the average annual growth of road transportation must be reduced to 4 percent. Assuming that the automobile utilization rate declines by 22 percent[7] from 2013 to 2030, the 4 percent growth in road transportation implies that the total number of passenger vehicles must be limited to 250 million. This means that the average annual growth in vehicle ownership must be substantially reduced from the 20 percent experienced during the 2008–2012 period to 6 percent from 2013 to 2030.

Our estimate is admittedly significantly lower than numerous other projections that expect the number of vehicles to reach nearly 400 million,[8] with some even predicting a peak of 750 million vehicles. Forecasters generally argue that these figures are realistic because even 400 million vehicles means a penetration rate of just under 30 percent (280 vehicles/1,000 people), far lower than the 63 percent in the United States and 46 percent in OECD countries.

However, based on our PM2.5 emissions model, if vehicle ownership reaches 400 million units, China cannot reach its 2030 PM2.5 target, even if maximum efforts are made in enforcing environmental measures such as reducing coal emissions, lowering individual vehicle emissions, and increasing fuel efficiency.

Therefore, based on our model simulation, the vehicle ownership penetration rate should reach just 180 vehicles/1,000 people (18 percent) by 2030, which is not particularly low relative to the stage of China's economic development. China's per capita GDP in 2030, calculated in constant US dollars, is estimated to reach the average level of OECD countries in the early 1970s. At that time, the average passenger vehicle penetration rate in OECD countries was only about 19 percent.[9] Furthermore, comparison between China and the United States on vehicle penetration is not ideal because America's population density is far lower than that of China and Europe, which means it has greater environmental capacity to accommodate a much larger vehicle fleet. In fact, it probably makes more sense for China to refer to Europe for its energy, environmental, and transportation models than to the United States.

Controlling passenger vehicle ownership to under 250 million does not imply that this is the peak, however. If China's air quality is immensely improved, and clean energy vehicle technology advances considerably, vehicle ownership will have room to grow. We do not rule out the possibility that the vehicle penetration rate in China could reach 400–500/1,000 between 2050 and 2060.

In addition to reducing air pollution, controlling the number of passenger vehicles will relieve urban traffic congestion and reduce energy consumption. Moreover, limiting the number of automobiles through measures such as an automobile license plate auction system and congestion fees will also provide local governments with new sources of fiscal revenue (see part two for more details).

1. Boost Rail Infrastructure

Rail transportation should be expanded at the same time that vehicle ownership growth is being limited. It is well known that rail travel generates far less air pollution and consumes less energy than road and air transport. The energy consumed by high-speed rail is equivalent to just one-twentieth to one-tenth of that consumed by highway and air travel, according to experts (see figure 3.4).

The pace of rail construction slowed between 2011 and 2013, following a tragic 2011 high-speed rail crash,[10] and due to a Ministry of Railways weighed down by debt. The government planned to increase total rail mileage from 90,000 km in 2011 to just 120,000 km in 2015 and has yet to propose a target for 2020, though according to discussions with experts, that target will likely be 140,000 km. Based on this estimate, by 2020, China's rail mileage per thousand people will still only be about 15 percent of the average length found in certain OECD countries, and far behind that of Canada (see figure 3.5).

Assuming road transportation growth of 4 percent as noted above, and total transportation volume growth of 5.5 percent, then rail/subway transportation (95 percent of which is rail) needs to average

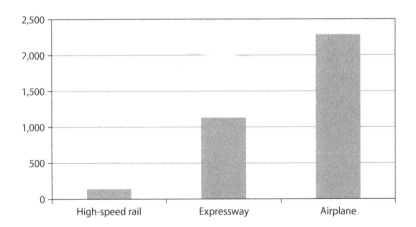

Figure 3.4 Energy consumption of various modes of transportation. *Source:* Zheng Qipu, "The Beijing-Shanghai High Speed Railway and Environmental Protection," *Journal of Railway Engineering,* March 1998.

Note: This shows an index of fuel consumption per passenger-km; ordinary rail is 100.

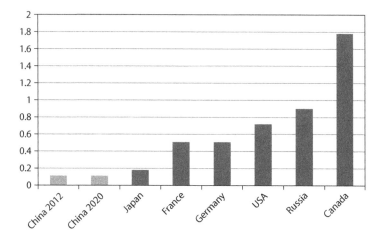

Figure 3.5 Railway density across countries (km/1,000 people). *Sources:* Author's estimates; *China Railway Yearbook*.

7.1 percent growth. Assuming further that the average railroad transportation efficiency for 2013–2030 is 1.2,[11] then total rail length needs to increase 165 percent during that period. This implies that by 2020, total rail length should reach 160,000 km, up 63 percent from 98,000 km in 2012 and 15 percent longer than the 140,000 km currently planned. By 2030, total rail mileage should reach 255,000 km, with annual average growth of 5.5 percent (see tables 3.6 and 3.7).

2. Expand Subways, Too

For the same transportation volume, the energy consumed and pollution generated by subways and urban light rail are 5–10 times lower

TABLE 3.6
Annual Average Growth in Transportation Volume, 2010–2030

GDP growth	6.8%
Growth in total transportation volume	5.5%
—Growth in total highway transportation volume	4.0%
—Growth in total railway/subway transportation volume	7.1%
—Growth in total railway transportation volume	6.6%
—Growth in total subway transportation volume	17.0%
Growth in railway length	5.5%

Source: Author's estimates.

TABLE 3.7
Forecasts of Total Railway Length

	Original forecast	Latest forecast
2020 (thousands of km)	140	160
2030 (thousands of km)	178	255

Source: Author's estimates.

than by private cars. At present, the government plans to increase the total subway and light-rail mileage (hereinafter referred to jointly as subways) from 2,000 km in 2012 to 7,000 km in 2020, which, in our view, is inadequate. At that pace, it will be impossible to satisfy the growing demand for public transportation that will result from limiting passenger vehicle ownership.

Based on our model, we recommend increasing the 2020 target by 40 percent to 10,000 km, and to 33,000 km by 2030. Only by substantially increasing subway length will it be possible to keep automobile traffic growth in major cities to about 4 percent. Our estimate is informed by traffic data for Beijing, Shanghai, and Wuhan: 46 percent of residents in those major cities choose private cars and taxis, 7 percent take subways and light rail, and 47 percent take buses.[12] Assuming urban transportation needs will grow 6 percent a year, and road transportation volume will grow 4 percent,[13] then average annual growth of urban public transportation must reach 9 percent.

Furthermore, considering that subways are more energy efficient, create less pollution, and have faster operating speeds than public buses, we project that the average growth in total subway mileage should be 17 percent and public bus transportation 6.5 percent between 2013 and 2030. Expanding total subway mileage at the rate above through 2030 will lead to subway density (calculated as total subway passenger-km/urban population) in major Chinese cities that is equivalent to about 75 percent of the subway density found in cities such as Singapore, Tokyo, and Hong Kong.

Because of the low-base effect and the relatively fast pace of urbanization, we project that total subway mileage will average growth of 22 percent from 2013 to 2020 and slow to 13 percent between 2020 and 2030 as the pace of urbanization declines. These rates of growth should lead to 10,000 km of total subway mileage by 2020 (five times the 2012 level)

TABLE 3.8

Growth of Various Modes of Transportation in Major Cities, 2013–2030

	2012	2013–30	2030
	As a proportion of overall transportation volume	Average growth rate	As a proportion of overall transportation volume
Private cars/taxis	46%	4.0%	26%
Public buses	47%	6.5%	41%
Subways	7%	17.0%	33%

Source: Author's estimates.

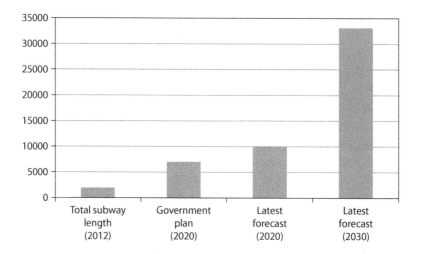

Figure 3.6 China's total subway length through 2030 (km). *Sources:* Author's estimates; China Association of Metros.

and 33,000 km in 2030. Our proposed subway length for 2020 is 40 percent higher than the government's current target, requiring additional growth of 230 percent from 2020 to 2030 (see table 3.8 and figure 3.6).

V. Structural Adjustment's Effect on PM2.5 Reductions

For ease of calculation, the sequencing of structural adjustments in our model was designed to address secondary industry first, reduce coal consumption second, and boost rail transportation last. Based on this sequence, the extent to which each structural adjustment effort will contribute to PM2.5 emissions reduction is roughly 6 µg/m³ from secondary industry, 8 µg/m³ from coal reduction, and 2 µg/m³ from rail transportation. If, assuming that the reduction target of 35 µg/m³ is achieved, then each factor contributes 18 percent, 24 percent, and 6 percent, respectively, to overall emissions reductions.

It is worth noting, however, that sequencing matters and has implications for the extent to which environmental measures and structural adjustment, respectively, can reduce emissions. Although the assumption in our model is to undertake structural adjustment after environmental measures have been maximally applied, it is worth discussing briefly the hypothetical results of reversing the sequencing of these actions.

First, the results from the structural adjustment above may be interpreted as less impressive because of the base effect. This is because if end-of-pipe control efforts are not fully realized, as is assumed in our model, then the total emissions base number will be greater initially. Therefore, implementing structural adjustments using a greater base of emissions volume will result in greater reduction of emissions. In fact, if we assume the sequence is reversed—that is, structural adjustments are executed before end-of-pipe control measures— then the magnitude of reductions from structural adjustment will be roughly three times larger than if it were pursued after environmental measures. More specifically, if the reverse sequence were assumed, then structural adjustments would reduce PM2.5 emissions by 61 µg/m³, far higher than if implemented after end-of-pipe measures (see table 3.9).

In reality, end-of-pipe control measures and structural adjustments will occur simultaneously, since many of the environmental actions have already begun. Therefore, the actual effects of the structural adjustments on emissions reduction should be somewhere between the two extremes noted above.

Second, the degree to which specific structural adjustment components contribute to PM2.5 reduction also depends on the sequence in which they are implemented. For example, if rail expansion is sequenced first, then its contribution will be greater than the current estimate (see table 3.10). This is because the last set of actions will be taken in the context of the smallest base number, so the absolute volume of emissions reduction resulting from those actions will be smaller.

TABLE 3.9
Effects of Structural Adjustments before Environmental Measures

	PM2.5 levels (μg/m³)	Emissions reductions (μg/m³)
No measures are taken by 2030	173	
Overall reductions from structural adjustment before end-of-pipe measures		61
— Nine percentage point reduction in secondary industries as a proportion of GDP	153	20
— Reduction in the proportion of conventional coal in total energy consumption from 67 percent to 33 percent	125	28
— Reduction of vehicle ownership from 400 million to 250 million vehicles	112	13
Overall emissions level from end-of-pipe control measures after structural reforms are completed	30	82

Source: Author's estimates based on the PM2.5 emissions control model.

TABLE 3.10
Effects of Structural Reforms after Environmental Measures

	PM2.5 levels (μg/m³)	Emissions reduction (μg/m³)	Contribution
Overall reduction from end-of-pipe control measures	46	19	54%
Structural adjustments after end-of-pipe control measures are complete		16	46%
— Nine percentage point reduction in secondary industries as a proportion of GDP	40	6	17%
— Reduction in the proportion of conventional coal in total energy consumption from 67 percent to 33 percent	32	8	23%
— Reduction of vehicle ownership from 400 million to 250 million vehicles	30	2	6%

Source: Author's estimates based on the PM2.5 emissions control model.

CHAPTER FOUR

Enabling Change:
Incentives Needed

IN ECONOMIC THEORY, the reason for excessive pollution—that which exceeds the level required for maximizing benefits to society—is that both demand for and supply of polluting products are too high, leading to a higher equilibrium output. Polluting products can be those that generate pollution during the production process—such as during thermal power generation or heavy industry production, among others—or during the consumption process, such as when we drive our cars. The cause of excessive output and consumption of polluting products has to do with the incentives in place. Conversely, inadequate output of less-polluting products, such as clean energy and subways, is also because of perverse incentives that lead to lower equilibrium demand and supply of such products.

So investors, manufacturers, and consumers need a new set of economic incentives to enable the proposed structural adjustments detailed in the previous chapter, which are necessary to meet the national PM2.5 reduction target. If implemented properly, these incentives and policies should be able to correct the tendency toward overexpansion of the secondary industry and encourage firms and individuals to reduce coal consumption, rely more on cleaner sources of energy, invest in pollution control technologies, reduce private car ownership, and use more public transportation.

In the first section of this chapter we will discuss the need to resolve three theoretical problems: market distortions (the state intervention problem), market failures (the market problem), and changing the objective function of firms and consumers (the market participants' problem). The goal of resolving these problems is to change the incentives for, and behaviors of, investors, firms, and consumers. In the following sections we will discuss nine types of economic policies that will enable the three types of structural adjustments—secondary industry, coal, and transport—to be successfully executed. These policies are concentrated in three areas: fiscal (spending, subsidies, and taxes), land (supply and pricing), and financial (new institutions and instruments for private capital allocation).

I. Market Distortions, Failures, and Participant Problems

1. State Intervention Problem

In the past few decades, the Chinese government has consistently supported exports to maintain employment and has pursued investment-driven stimulus policies to sustain economic growth. And because the Chinese Communist Party (CCP) cadre performance review system puts a premium on growth, local officials were incentivized to bolster GDP— the most direct means of doing so was attracting business and capital on a large scale. To a great extent, these strategies led to the extraordinary development of real estate and manufacturing. Consequently, tax and land policies were biased toward high-polluting manufacturing and construction industries and crowded out the low-pollution services sector.

For example, three types of policies artificially stimulated the manufacturing sector. First, local governments provided abundant industrial land, resulting in industrial land prices that are only one-eighth the price of residential land—a ratio that is 1–2 times lower than the average in other countries. Second, the indirect tax burden on many services businesses is heavier than the indirect tax burden on manufacturers. While these policies served a positive role during one stage of China's economic development, they have now clearly become distortions that create bottlenecks in economic structural adjustment.

Finally, the bias toward an expansive manufacturing sector is also related to high entry barriers for services firms. For example, sectors such as financial services, telecommunications, healthcare, education, railroads, subways, civil aviation, and the media are highly regulated by the government, making it difficult for private and foreign capital to gain access, which results in insufficient supply. Such artificially and policy-induced distortions need to be untangled through reforms, in particular by opening up access to broader competition.

2. Market Failures

The unbalanced nature of China's energy and transport systems—very high levels of coal consumption coupled with low levels of clean energy consumption and rapid growth in automobile traffic—are symptoms of market failures.

Take coal consumption for example. If the price of coal were determined solely by market supply and demand, it would result in the overproduction and consumption of coal, generating significant pollution. That's because the "negative externality"—the pollution and health hazards posed to others—resulting from coal consumption is not priced into the transaction between producers and consumers, who prioritize maximizing profits (producers) and utility (consumers) in a pure marketplace. This leads to a lower price for coal than it would otherwise be if the negative externality had been accounted for via a tax on the coal price.

So in order to address the problem of low coal prices, the government can use taxes, such as increasing the resource tax or levying fees on coal-related pollution, to raise the cost for producers (or the price for consumers), thereby curtailing production and consumption. The failure of the market to fully price in the negative externality of coal consumption is the main reason why China consumes so much of it.

Inadequate investment in clean energy is another example of a market failure. Given the same energy output, when compared to coal, clean energy can reduce SO_2 and NOx emissions by at least 90 percent. In addition, clean energy has "positive externalities"—that is, the beneficiaries of emissions reduction are not primarily the producers or consumers of clean energy, but rather other residents who breathe the local air. In this sense, clean energy functions as a public good because it enables the broader public to live healthier lives, thereby reducing future medical expenditures and boosting productivity.

But because the beneficiaries are not paying for the positive externalities of the clean energy project, the result is a relatively lower profitability for clean energy producers. Since producers find it difficult to turn a profit, their willingness to invest in this sector is weak. Although the government already provides partial subsidies, those subsidies are insufficient, and the positive externalities cannot be fully endogenized, resulting in a persistently low proportion of clean energy relative to total energy consumption in China.

Finally, the rapid growth in automobile consumption is another example of failure of the market to price in negative externalities. Fully market-determined vehicle prices would mean lower cost for automobiles, resulting in excessive consumption. That's because neither the producers nor the buyers take into account the pollution and traffic congestion costs impose on others. ("Excessive consumption" is used here in relation to the overall well-being of a city that requires clean air and traffic-free roads, and not to the utility of automobile producers and buyers.)

In order to correct this market failure, an astute government would manage excessive consumption by increasing the cost of buying an automobile through license plate fees or congestion fees. The rapid growth of automobiles in many major Chinese cities, such as Beijing and Chengdu, as well as the resulting pollution and congestion, primarily reflect the policymakers' failure to fully recognize the negative externalities of automobile consumption and the need to use policy instruments to curtail consumption. For many cities where the automobile penetration rate is already approaching 20 percent (200 vehicles/1,000 people), measures to limit car ownership must be taken in the next several years or the consequences for urban pollution and traffic may become too awful to contemplate.

3. Businesses and Consumers Lack a Sense of Social Responsibility

As market participants, Chinese firms and consumers have assigned very little weight to social responsibility in their objective functions, which has exacerbated the challenges of deploying clean energy and transforming modes of transportation.

First, the classic assumption in microeconomics is that firms only pursue profit maximization and eschew corporate social responsibility (CSR). Based on a given output price and input costs, firms arrive at an optimal output quantity that will maximize their profits. But in fact, firms can pursue dual objectives that comprise both profits and CSR, as experiences in developed markets have demonstrated. A socially responsible firm's profits can be positively correlated with its output of clean products and negatively correlated with its output of polluting products.

The integration of social responsibility into businesses' objective functions has already manifested itself in the public disclosures of major financial institutions and publicly listed companies in certain developed economies. For example, Germany's Deutsche Bank 2012 CSR report states, "We believe our responsibility goes beyond our core business. Progress and prosperity are driving us when we initiate and support educational, social, and cultural projects."[1] US-based General Electric also states, "At GE, sustainability means aligning our business strategy to meet societal needs, while minimizing environmental impact and advancing social development. This commitment is embedded at every level of our company—from high-visibility initiatives such as Ecomagination and Healthymagination to day-to-day safety and compliance management around the world."[2]

To a certain extent, if the weight assigned to CSR among a corporation's objectives is greater than zero, it can substitute for fiscal and other policies to incentivize changes in corporate behavior. Specifically, the CSR weighting can be increased to achieve the same effect as government subsidies for clean products. A system can be established for weighting CSR in a firm's objective functions, through mechanisms such as disclosure requirements, green credit rating, environmental liability, and public education (see chapter 9 for more details).

When it comes to consumers, the classic assumption in microeconomics is that consumers only pursue maximization of utility. In other words, market prices are affected by consumer preferences, which also determine the magnitude of externalities. But consumers can also develop a sense of social responsibility—that is, for responsible consumers, a product's price and utility may not be the only factors in their purchasing decision.

Once again, as experience in developed countries can attest, many consumers have begun to incorporate moral values and a sense of responsibility in their consumption behavior. These "values-based" consumers want to know how a product is produced, where it is produced, and even the factory it is produced in, as well as whether the factory engages in malpractices such as polluting the environment, employing child labor, or misappropriating intellectual property, among others. These consumers are often willing to pay a price premium for products that align with their values. For instance, a 2011 Gallup poll conducted with 31,000 consumers in 26 countries showed that 50 percent of those surveyed were willing to pay more for products manufactured using clean energy.[3]

If social responsibility in a consumer's objective function is weighted greater than zero, then it can be proven that the consumer's optimal choice is to consume more clean products than if social responsibility carries zero weight. In other words, the demand for clean products among socially responsible consumers is greater than the demand for clean products among consumers who are not socially responsible.

Once consumers have developed a sense of social responsibility, their demand for clean products should increase, which in turn causes the price of clean products to rise, thus incentivizing producers to provide more supply. This is an effect similar to government subsidies for clean products. There are various ways to increase consumers' sense of social responsibility, including providing environmental education to children, disclosing the facts about firms' environmental performance, creating environmental role models, and using public opinion to censure certain consumer behaviors, among others.

To summarize, economic policies that enable structural adjustments should aim to achieve three broad objectives: correcting inappropriate state interventions in the market, addressing market failures (externalities), and increasing the weighting assigned to social responsibility in the objective functions of firms and consumers. In the following sections we will discuss nine specific measures and policy actions for meeting these three objectives.

II. Policies and Actions to Incentivize Structural Adjustment

1. Reform Industrial Land Prices

At the end of 2012, the average price of residential land in 105 Chinese cities was 4,620 yuan ($745)/m², while industrial land was 670 yuan ($108)/m², just about 14 percent of the residential land price—a ratio that makes China an outlier when compared to other countries and regions. For instance, the industrial to residential land price ratio is approximately 60 percent in Taipei, Taiwan;[4] in Tokyo, that ratio was roughly 100 percent in 2012. In other areas of Japan, the ratios are roughly 76%, such as Tokyo-to, Chiba-ken, Osaka-fu, and Kumamoto-ken.[5] In the United Kingdom, the industrial to residential land price ratio is about twice that of China.

One important cause of cheap industrial land can be attributed to the irrational allocation of land supply at the local level. Across 40 medium and large cities in China, industrial land accounts for 40 to 50 percent of total land supply, while residential land accounts for only around 20 percent. This is again in contrast to other countries, where industrial land supply generally does not exceed 15 percent and residential land typically accounts for around 45 percent.[6]

Ultimately, cheap industrial land, often provided through tax rebates and other local government subsidies, is a distortion of the market that has contributed to the disproportionate expansion of industry in the Chinese economy, contributing to pollution. Reforming industrial land price and allocation requires several different approaches.

First, at the national and provincial levels, industrial land supply must be significantly reduced while residential and commercial land supply is bolstered. For existing industrial land, rezoning should be permitted. For example, the "Seven Baselines" published by the Beijing Municipal Construction Committee stipulates: "Upon conforming to the relevant legal and regulatory provisions and planning require- ments, business and public institutions and related organizations are encouraged to use their own land for the development and construc- tion of owner-occupied commodity housing." Because industrial land prices are far below residential land prices, this measure will result in the conversion of industrial land for residential purposes. Such a conversion process will indirectly cut the supply of industrial land and naturally increase its price, while it will simultaneously lower the cost of residential and commercial property, tamping down housing prices and encouraging the development of the services sector.

Second, land resource administrators at the provincial level should raise the minimum price for the transfer of industrial land, and dis- close the minimum prices for different industries and regions. Finally, because industrial firms often engage in hoarding low-cost land, those that fail to use the land for the purposes stipulated in the contract must be regulated, which will effectively serve to increase the cost of hoarding industrial land.

2. Lower Tax Burden on Services

According to a number of studies, the high tax burden on the services sector effectively encourages overinvestment in the industrial sector.

For instance, a quantitative analysis by Ping Xinqiao (2011) indicated that in eleven of the relatively more developed coastal provinces and municipalities, services such as logistics, IT and software, real estate, leasing and financial, and R&D are all subject to higher indirect tax burdens (indirect taxes as percent of total value added) than are industrial firms above a certain scale.[7]

In addition, Luo Minghua (2010) compared the respective tax revenues and GDP for the services and manufacturing sectors for 2005–2008, concluding that the tax burden—tax revenue as a proportion of industry value added, including direct and indirect taxes—for the services industry has continuously increased relative to the manufacturing sector, and has surpassed it since 2007.[8] Zhang Lunjun and Li Shuping (2012), too, have compared the overall tax burdens on secondary and tertiary industries (including income taxes), similarly concluding that when direct taxes are included, the tax burden on tertiary industry grew rapidly, exceeding that of secondary industry in 2007.[9] Finally, Tang Dengshan and Zhou Quanlin (2012) used input-output table data to extrapolate VAT equivalents for the various industries subject to a business tax and found that the tax burdens on the construction, hotel, and food services industries were notably higher than the average for all industries.[10]

Many developed countries adopt a fairly neutral tax system that neither offers preferential taxes nor discriminates against the services or the manufacturing sectors. In countries such as Australia and New Zealand, certain services are entitled to preferential taxes, such as education, healthcare, transportation, financial services, and insurance, so that the tax burden on the services sector is actually lower than that of the manufacturing sector.

To address the tax burden issue in China, certain pilot programs are underway to convert the business tax to a VAT, though the results may be unexpected. That's because since the manufacturing sector has already been subject to a VAT, once the services sector converts from the business tax to VAT, manufacturing firms will be able to obtain even more invoices for VAT refunds, which actually further reduces their tax burden. In some cases, the decrease in the tax burden on manufacturing is even greater than that for services.

After initiation of the business tax-to-VAT conversion pilot program in Liaoning province, the overall tax reduction for thirteen cities across the province (excluding Dalian) is projected to be about 2.2 billion yuan ($355 million).[11] Deductible input VAT for ordinary taxpayers is expected to increase by over 1.3 billion yuan ($210 million), and secondary industries stand to benefit. Judging from data published after the five-month pilot program to convert business tax to VAT in Jiangsu province, the manufacturing sector in this province experienced the biggest tax cut. During the pilot program, the ten industries with the highest VAT reductions in Jiangsu included the wholesale, transport, chemical raw materials, and chemical products manufacturing industry, with VAT reductions accounting for 59.8 percent of total reductions.[12] Of course, the data above only reflect a fraction of the pilot programs and are thus insufficient to accurately assess the relative changes in tax burdens. But the risk remains that tax reductions will actually benefit manufacturing more than services.

Currently, pilot programs to convert from the business tax to VAT are being fine-tuned and gradually expanded, and policies have yet to be fully determined for many new services industries. Before this reform is implemented across the board, and particularly before new VAT rates for services industries are determined, we recommend careful consideration of input VAT deductions for the various services and manufacturing industries in order to avoid indirect tax burden bias against services.

3. Raise the Resource Tax Rate on Coal

Improving the energy structure by reducing the proportion of conventional coal consumption should primarily rely on economic incentives, such as substantially increasing the relevant taxes and fees for coal and significantly boosting subsidies for clean energy. The focus here is on the coal resource tax.

Until December 2014, China maintained a volume-based coal resource tax that was virtually meaningless in curtailing coal consumption. For instance, in major coal producing regions, volume-based resource taxes were collected at the rate of 3–5 yuan ($0.48–$0.80)/ton, equivalent to just 0.5–0.8 percent of the production cost of coal. To control demand for coal, many experts, ourselves included, recommend adopting a value-based (*ad valorem*) coal resource tax of 5 percent, or 5–9 times higher than the previous levels.

According to coal economist Jiao Jianling's research, the demand price elasticity of coal is 0.96. Based on this elasticity and the calculations of our CGE model, if the coal resource tax is increased to 5 percent in tandem with higher pollution fees and subsidies for clean energy, it should be possible to reduce coal demand by approximately 4 percent compared with the baseline (no reform scenario) during the period of the policy implementation.[13] This rate of reduction aligns with our forecast of slashing conventional coal consumption by 14 percent through 2030. In fact, the State Council in late 2014 decided to implement the ad valorem resource tax, permitting each province to determine its rate within the 2–10 percent range. The rates already announced by Henan, Hunan, Guangxi, Shanxi, Shaanxi, and Inner Mongolia were 2 percent, 2.5 percent, 2.5 percent, 8 percent, 6 percent, and 9 percent, respectively—an average of nearly the 5 percent we proposed.

While the imposition of an ad valorem resource tax on coal is a step in the right direction, we believe the average resource tax should continue to increase over the longer term, perhaps to as high as 10 percent, to further raise the cost of coal and incentivize a transition to cleaner energy sources.

4. Levy Higher Pollution Fees and Impose a Carbon Tax

Until 2013, the levies on SO_2 and NOx emissions were too low, especially when compared to those of developed economies (see table 4.1). For example, the pollution charge for SO_2 reached 1.20 yuan ($0.19)/kg in only two provinces in China in 2013, while a survey of multiple power plants showed that the unit desulfurization cost at thermal power plants was as high as 3 yuan ($0.48)/kg[14] and desulfurization costs were even higher for other companies because they did not enjoy electricity price subsidies. For NOx emissions, too, the pollution fees were generally below 1 yuan ($0.16)/kg, and were raised to 1.26 yuan ($0.20)/kg in early 2013 in only a few cities, such as Shanghai. Because these fees were much lower than the cost of installing emissions reduction equipment, many companies simply opted to pay the pollution charges rather than installing desulfurization and denitration facilities.

Numerous experts have recommended increasing pollution fees for SO_2 and NOx. In addition, we also recommend correspondingly increase pollution levies on soot emissions, sulfuric acid mist emissions, and dust emissions. In early 2014, many cities and provinces, including Beijing and Tianjin, began to aggressively raise the SO_2 and NOx emissions charges to 5–10 yuan ($0.81–$1.61)/kg.

TABLE 4.1
Comparison of Pollution Levies Across Countries (yuan/kg)

	SO$_2$ emissions	NOx emissions
Sweden	13.4	46.8
Denmark	12.4	28
Norway	12.4	16.5
Italy	0.9	1.8
Australia*	N/A	1.7 8.9
Estonia	0.7	0.9–16.7
Hungary	N/A	3.6
China (2012)	0.63–1.26	<1

Sources: OECD; finance ministries of various countries.
*Australian data is for New South Wales.

A carbon tax, which is generally considered a climate change action to cut CO$_2$ emissions, can also have an effect on PM2.5 emissions. That's because for many fossil fuels, especially coal, SO$_2$, and NOx, emissions (main sources of PM2.5) are highly correlated with CO$_2$ emissions (not part of PM2.5). Therefore, a carbon tax can simultaneously help reduce PM2.5 emissions and curb CO$_2$ emissions.

According to the Chinese Academy for Environmental Planning project team, carbon taxes of 11 yuan ($1.77), 17 yuan ($2.74), and 12 yuan ($1.94) should be collected per ton of coal, ton of petroleum, and thousand cubic meters of natural gas, respectively. Based on a coal price of 600 yuan ($97)/ton, oil price of 4,200 yuan ($677)/ton, and natural gas price of 3 yuan ($0.48)/m³, this is equivalent to collecting an ad valorem carbon tax of 1.8 percent for coal and 0.4 percent for petroleum and natural gas. We believe such a tax should be introduced in the next few years.

5. *Substantially Increase Subsidies for Clean Energy*

At present, the Chinese government's subsidies for clean energy account for 0.2 percent of fiscal expenditure, while the same figures in the United States and Germany are 0.4 percent and 0.7 percent, respectively (see fig. 4.1). As discussed above, our simulation suggests that it is necessary for clean energy as a proportion of total energy

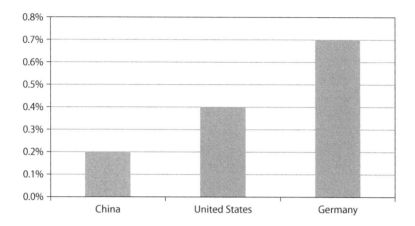

Figure 4.1 Clean energy subsidies as a proportion of total fiscal spending, 2011–2012. *Sources:* WIND; Ministry of Finance; *21th Century Business Herald.*

consumption to nearly double from 14.6 percent to 27 percent from 2013 to 2020. This means that fiscal subsidies on clean energies must also correspondingly increase.

Based on our estimated growth rate for clean energy and the per-kilowatt subsidies currently required, fiscal subsidies for clean energy as a proportion of total fiscal spending should rise to 0.4 percent in the coming years. Increasing spending in this area will not necessarily lead to larger fiscal deficits because new sources of revenue—such as the resource tax and pollution fees, as well as the introduction of the carbon tax and revenue from automobile license plate auctions—could help to offset the additional fiscal spending required.

6. Improve License Plate Auction System and Impose Congestion Fees

Global experiences suggest that license plate auction systems, congestion fees, and other taxes or charges related to automobiles are effective ways to reduce car ownership and their usage rate. These two measures can be applied separately, or they can be used simultaneously. This section looks at how these measures can be applied to China.

I. LICENSE PLATE AUCTION In our view, major Chinese cities should adopt Singapore's automobile license plate auction system to significantly cut annual average passenger vehicle ownership growth from 20 percent in the years before 2013 to around 6 percent from 2013 to 2030.

Currently, Singapore's certificate of entitlement system effectively limits growth in vehicle ownership to 0.5 percent annually, despite annual population growth as high as 3 percent in recent years, as the government encourages the population to grow from 5.3 million to 6.9 million by 2030. Singapore's certificate of entitlement system, in effect, makes a car cost 400 percent more than the actual sticker price because of the additional taxes and fees required to own a private automobile. In Shanghai and Beijing, the taxes and fees for owning a private car are about 110 percent and 35 percent of the cost of the car, respectively (see fig. 4.2).

In Shanghai, although a license plate auction system already exists, the number of license plates available each year, about 200,000, was still too high as of 2013. This implied 10 percent annual growth in private car ownership, which would exacerbate pollution and congestion. We recommend that Shanghai substantially limit license plate availability each year to lower vehicle ownership growth to 2 percent. Moreover, Shanghai needs to change the license plate system from indefinite duration of validity to an expiration system—perhaps a validity period of ten years. This is one way to spread the costs of maintaining the

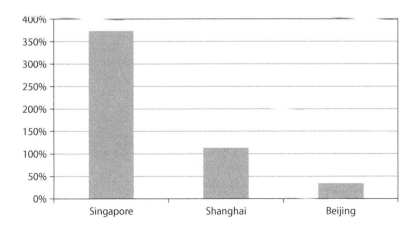

Figure 4.2 Passenger vehicle taxes and fees as proportion of purchase price. *Sources:* WIND; General Administration of Customs; *Xinhua.*

system so that they are borne by all "polluters," not just by consumers who buy new cars. The advantage of this system is that a large number of license plates will expire each year. This means that given the 2 percent annual growth in vehicle ownership, additional new license plates can be auctioned off, which can alleviate the supply shortage and prevent the price for new license plates from soaring.

Of course, a sudden policy change will invariably create tensions with owners of older license plates who assumed they would remain "permanently valid." To deal with this, the Shanghai government can compensate holders of older license plates by modestly extending the valid term of plates, perhaps slightly longer than ten years, while simultaneously moving to an expiration system (see chapter 6 for a case study of Shanghai).

Beijing, in contrast, uses a lottery system to allocate of license plates, which has obvious disadvantages when compared to a license plate auction system. First, compared to the auction system, the lottery system allocates license plates to persons with lower marginal utility, resulting in a significant loss of consumer utility (social welfare). That's because the lottery system offers all participants the same probability of obtaining a license plate, irrespective of the marginal utility of owning a license plate.[15] In other words, the people who win the lotteries are not necessarily those who are most in need of the license plate (defined as the persons having the highest marginal utility).

We estimate that this loss in consumer utility is as much as 12 billion yuan ($1.9 billion) annually. To illustrate, assume that of the 1 million people who participate in a monthly lottery, 20,000 win license plates. Furthermore, assume a linear distribution of their marginal utility of winning a license plate from virtually zero to approximately 100,000 yuan ($16,129 about the price in Shanghai auctions). Finally, assume that this marginal utility is randomly distributed among lottery winners—that is, the average utility per winner is 50,000 yuan ($8,065). Thus, on average, under the lottery system, the marginal utility of each person who wins a license plate is 50,000 yuan ($8,065) less than the marginal utility of a license plate obtained under the auction system. The total utility loss for each auction (20,000 license plates) is therefore 1 billion yuan ($161 million), and the total utility loss for the twelve auctions held each year is 12 billion yuan (nearly $2 billion).

Second, compared to auctions, the lottery system causes local governments to lose an important source of fiscal revenue. For example,

Shanghai's license plate auctions in 2012 generated 7.1 billion yuan ($1.1 billion) in local revenue. If Beijing had that additional revenue, it could have accelerated the construction of subways and clean energy projects (see chapter 7 for a case study of Beijing).

II. CONGESTION FEES Another policy option to control traffic and car-related pollution in major cities is to impose congestion fees, which are used in cities such as London, Florence, and Singapore. The basic idea behind imposing a congestion fee is to charge car users in areas that are afflicted with bad traffic and environmental stress and direct non-essential drivers to switch to public transportation. With the exception of public vehicles such as buses, ambulances, fire engines, and police cars, all automotive vehicles entering designated areas are required to pay the congestion fee. In the case of London, starting in 2003, the city charged vehicles entering the city center a congestion fee of £5 ($7.50) per day. The policy did indeed reduce traffic volume by 12 percent, consistent with the London government's predictions based on price elasticity. The fee has been progressively increased to the current £10 ($15) per day, in keeping with inflation.

Auctioning license plates and introducing congestion fees are two options for controlling congestion and pollution in China's major cities. They can be used individually or in combination. If the license plate auction system is already effective in controlling vehicle ownership growth, then it may not be necessary to impose additional congestion fees. On the other hand, if a government feels there is too much public resistance to a license plate auction system, then it may opt for a congestion fee, which tends to be less politically controversial.

7. Reform the Financing System for Rail Transportation

As discussed above, aggressive expansion of rail infrastructure, including railroads and subways to increase the proportion of urban residential travel by subway, should be a key option for reducing transportation-related pollution. However, funding is a major bottleneck in expanding rail transportation. China Railway Corporation, one of the major state-owned rail and subway owners and operators, has accumulated debt of nearly 3 trillion yuan ($484 billion) and is facing difficulties in mobilizing new financing for major projects. As China's growth slows, local fiscal revenue growth is also decelerating, while local government financing platforms are running into problems in obtaining outside financing because of high debt ratios and risk. Consequently, it will be very difficult to sustain the 20 percent annual growth in subway investment that is needed to meet the projected target of 10,000 km of subways by 2020.

Therefore, the local financing model for subway transportation must be reformed to enable changes in the transportation structure. Specifically, reforms should be focused in three areas: (1) attracting private capital for subway investment, (2) implementing license plate auction and congestion fees, (3) issuing municipal bonds.

In terms of attracting private capital, Japan and Hong Kong both offer good examples to follow. Beginning in 1987, Japan restructured the heavily indebted Japanese National Railways, splitting it into six regional passenger transportation companies and one cargo transportation company. These new entities subsequently underwent initial public offerings (IPOs) to achieve partial privatization. After the restructuring process, efficiency, investment, and profitability all increased, while debt and government subsidies declined by 80 percent.

In Hong Kong's case, the municipal government owned the city's Mass Transit Railway (MTR) when it was founded in 1975. After its IPO in 2000, the government continued to hold a 77 percent controlling share in MTR, and the public held the rest. While MTR fares are quite affordable, it has managed to become one of the few profitable subway companies in the world. A key to the company's success lies in its use of the "subway + land development" model. The municipal government sells highly profitable land zoned for residential and commercial use to the subway company, which is also a real estate developer, together with the less profitable subway construction rights, enabling the government to attract private investment into the projects without using fiscal subsidies. It also keeps subways and residential areas near each other. For the MTR, once the subway construction is finished, the price of residential units it builds rises accordingly.

This can be considered a form of public-private partnership (PPP) financing for green infrastructure projects. Major cities in China can draw on the experiences of Japan and Hong Kong by using IPOs and the PPP model to both raise capital and leverage private capital. Local governments in China can also consider directly attracting institutional investors, such as insurance companies and pension funds, to become shareholders in subway companies.

Automobile license plate auctions and congestion fees can provide an important source of funding for subway construction. As noted above, Shanghai's license plate auctions can provide the municipal government with 6–7 billion yuan (~$1 billion) in revenue each year. If the license plate auction system is reformed according to our recommendations and a new congestion fee is implemented, net fiscal revenue could be as much as 10 billion yuan ($1.6 billion), a substantial portion of which could be allocated to finance subway expansion.

Finally, municipal bonds can also be used to finance rail projects. A transparent bond market can provide a stable and long-term source of funding for rail projects (e.g., financing with maturities of ten to thirty years). While the Chinese government pledged to develop a local bond market at the November 2013 Third Plenum, major cities such as Shanghai and Beijing should lead the reforms. These would involve publishing balance sheets, increasing fiscal transparency, and obtaining credible municipal bond ratings, among others. Not only will this help finance rail infrastructure, it will also promote the development of the local bond market and gradually mitigate the financial risk of maturity mismatch (short-term borrowings for long-term investment) faced by many local financing platforms.

8. Create Interregional Compensation Mechanisms for Emissions Reduction

While China has explored interregional environmental compensation schemes over the last two decades, major pilot programs have not included air pollution but have instead covered water resource protection, forest preservation, erosion control, desertification prevention, and conversion of farmland to forests and grazing land to grassland.

Given the reality that interregional externality of air pollution is pronounced, the principle of interregional "joint prevention and control" has been put forward. In our view, there is an urgent need to establish interregional compensation schemes for PM2.5 emissions reduction. For example, about one-quarter of the PM2.5 emissions from one locality can drift 200 to 300 km,[16] an assessment consistent with the estimates of Ben-Jei Tsuang at the National Chung Hsing University in Taiwan.[17]

The PM2.5 interregional compensation scheme we propose is centered on the "beneficiary pays" principle. That is, funding is redistributed from developed areas (the primary beneficiaries) to help underdeveloped neighboring areas reduce atmospheric pollution. This reduces the spillover effects (externality) from under-developed areas' atmospheric pollution to developed areas, and achieve two objectives: (1) cut the cost of emissions reduction in the developed areas and (2) assist the underdeveloped areas with emissions control.

For example, about 40 percent of the PM2.5 sources in Beijing originate from peripheral areas, mainly Hebei province. Meanwhile, per capita income in Beijing is more than double that of Hebei, and the installation and utilization rates of industrial desulfurization equipment are far higher in Beijing than in Hebei. Going forward, while Beijing will need to rely on higher-cost measures such as deploying more clean energy, limiting automobiles, investing in subways, and subsidizing EVs to address PM2.5 pollution, there are still plenty of low-cost actions that can be taken in Hebei to mitigate pollution. Such measures include ramping up installation and utilization rates of desulfurization and denitration facilities, increasing natural gas power generation, applying energy conservation technology in heavy industry, and controlling biomass burning in rural areas, among others. Still, the Hebei government does not have sufficient funds to undertake these low-cost actions because it is a relatively poor province.

So even as Beijing plans to invest 1 trillion yuan ($161 billion) to mitigate atmospheric pollution, primarily in high-cost emissions reduction projects, the lack of interregional compensation mechanisms means the effects could be limited. That's because without concomitant actions taken in Hebei, the interregional externality of air pollution will still significantly affect Beijing.

Therefore, one Pareto-optimal economic policy should be to establish a Beijing-Hebei interregional compensation mechanism for air pollution control. We recommend that the Beijing and Hebei governments solicit expert proposals to vet a set of projects in Hebei that have the highest cost-benefit ratio. That is, these projects should be able to demonstrate that with Beijing's investment in Hebei, PM2.5 reduction outcomes will exceed the results Beijing would achieve if it were to invest the same amount in its own local emissions-reduction projects.

To illustrate, suppose that Beijing invests 100 billion yuan ($16.1 billion) in Hebei projects to reduce the province's PM2.5 emissions by 30 percent and thereby reduce Beijing's PM2.5 level by 10 percent, but that same investment in local projects can cut Beijing's PM2.5

emissions by only 5 percent. In this case, Beijing has sufficient reason to allocate the investment for Hebei projects and thereby obtain superior results.

So the design of any compensation mechanism should incorporate a list of emissions reduction projects in Hebei that have the highest cost-benefit ratios, thus incentivizing Beijing to invest or participate in the projects. Funding for this compensation scheme could come from several sources. It could come directly from Beijing's fiscal revenues, which could be used to subsidize desulfurization and denitration in Hebei's power generation and heavy industries. It could also come from Beijing's electricity price subsidies for consumers, to subsidize clean energy projects in Hebei, which in turn will be used to supply Beijing. Finally, it could come from fees paid by Beijing-based companies to Hebei firms in an emissions trading scheme.

9. Establish a Policy Framework for Green Finance

Creating a policy framework to direct social (private) capital into green industries will also be crucial in enabling the three necessary types of structural adjustments. We estimate that over the next several years, China will need to invest more than 2 trillion yuan ($323 billion) per year in green industries, in areas such as environmental remediation, energy efficiency, clean energy, and clean transportation. But the current financial system and market prices remain woefully inadequate to offer incentives for such large-scale green investment.

Our research demonstrates that four categories of policies can further incentivize green investing. The first involves policies that increase rates of return and lower the cost of investing in green projects. The second involves policies to lower the rates of return on polluting projects. The third includes policies that alter the objective function of firms from simply maximizing profits to the simultaneous pursuit of profits and social responsibility. The fourth comprises policies that shift consumer preference toward green products. Next we detail specific recommendations for establishing a green financing framework for China based on theory and international experience.

I. CREATE A NATIONAL-LEVEL GREEN BANK With the government as a founder, the national green bank should specialize in green lending and investing, leverage other financing sources such as green bonds, and fully realize the economies of scale of specialized project evaluation capacity. More specifically:

- Capital for the green bank can come from the government initially, and private capital can also participate as shareholders. Sources of private capital could include social security funds, insurance companies, and other institutions with long-term investment interests. Another primary source of funding would be the issuance of medium- and long-term green bonds.
- The bank will focus on financing specific industries and projects, primarily in the areas of large-scale environmental protection, energy efficiency, new energy, and clean transportation projects. In addition to green lending, the bank should also engage in activities such as green equity investing and guarantees.
- The bank should fully realize the leverage effect of government funding by attracting private capital, bond issuance, guarantees, and joint venture equity investing. It should strive to leverage 5–10 yuan ($0.81–$1.61) of private capital for every 1 yuan ($0.16) of state capital.
- The bank should assemble the best professional talent, information and analytical capacity to conduct environmental cost and benefit calculations, project evaluation, and risk control. By having this professional expertise within one bank, it can achieve the greatest economies of scale.

II. DEVELOP A GREEN BOND MARKET China should develop its own domestic green bond market, to help finance long-term green projects. If well designed, the green bond market has the potential to fund 20–30 percent of needed green investments in the country. The following steps should be taken to build a domestic green bond market:

- Regulators issuing green bond guidelines that specify the basic requirements for issuers, including information disclosure on the use of proceeds.
- Clearly defining the scope of "green bonds" by issuing a Green Bond Catalog.
- Facilitating the development of second-opinion or third-party verification service providers.

- Incubating institutional investors by increasing their preference for green assets, including green bonds.
- Exploring policy options to incentivize the issuance of green bonds, including interest subsidies and credit guarantees.

III. EXPAND COVERAGE OF INTEREST SUBSIDY PROGRAMS FOR GREEN PROJECTS

- Expand the list of projects that qualify for interest subsidies by incorporating industries such as clean energy, clean coal, and natural gas vehicles.
- Lift the cap on business size to allow more small- and medium-sized companies to qualify for interest subsidies.
- Remove the duration on interest subsidies (current cap is three years) and the cap on the subsidies interest rate (currently at 3 percent).
- Encourage local public finance bureaus to expand their interest subsidy program to local green projects.

IV. INCORPORATE ENVIRONMENTAL RISK INTO PROJECT EVALUATIONS AND ESTABLISH A GREEN CREDIT RATING SYSTEM

- In evaluating projects and borrowers, banks should incorporate environmental risk factors. For example, if a project involves air pollution, water pollution, solid waste, or other risks, a quantitative impact assessment report should be required, expert opinions should be solicited concerning potential policy changes, and reputational and legal risks the project or firm could face in the future should be outlined. For industries with high environmental risks, such as mining, power, forestry, fishery, waste disposal, petroleum and natural gas, metallurgy, and chemicals, impact assessments should be mandatory.
- Banks must disclose the relevant information for all green loans and loans to environmentally risky sectors in their annual reports.
- Banks should refer to the "Equator Principle" in establishing dedicated environmental and social risk control departments and adopt full-process management of projects that have an impact on the environment.

- Banks and rating agencies should establish a green credit rating system that incorporates the environmental risks for borrowers and bond issuers in credit rating reports.

Other green credit innovations could include:

- Lengthening green loan maturities by funding them with green bonds.
- Extending green loans to projects collateralized based on rights to collect fees (e.g., on water treatment or expected sales of carbon credit).
- Securitizing green loans for which underlying projects have stable cash flows.
- Offering preferential green loans, such as those for clean energy, environmental protection equipment, green construction, and new energy vehicles, among others.

V. DEVELOP A PUBLIC ENVIRONMENTAL IMPACT ASSESSMENT SYSTEM As they evaluate the environmental impact of investment projects, banks, institutional investors, and firms should be able to quantify emissions and other environmental costs, such as CO_2, SO_2, NOx emissions, wastewater discharges, and solid waste production, among others. The evaluation should determine the extent to which these costs ("externalities") are not reflected in market prices. Such a system could bring enormous social benefits, so we recommend that the government or a green investors network with official support take the lead on developing such a system. The final product, including the methodology, IT system, and database, should be made available to the public and to investors as a quasi-public good at the lowest possible cost.

VI. IMPLEMENT MANDATORY GREEN INSURANCE IN MORE SECTORS The MEP and the China Insurance Regulatory Commission (CIRC) jointly published the "Guiding Opinion on Launching Work on a Pilot Program of Mandatory Pollution Liability Insurance" in early 2013[18] that proposed a number of areas that must be covered by green insurance. However, the scope of industries covered in the opinion is limited, and focuses primarily on water pollution and soil contamination. We recommend an expansion of the industries covered by green insurance.

In addition to the maritime oil exploration and river transportation industries, which are already required to participate in green insurance, petroleum and natural gas extraction, petrochemical, thermal power, coal mining and processing, chemical coal processing, steel, cement, plastics, and hazardous chemicals transportation and processing should also be covered under the mandatory insurance scheme. Once revised, this opinion should become a full-fledged regulation and should be implemented as soon as possible.

VII. MANDATE ENVIRONMENTAL INFORMATION DISCLO-SURE BY LISTED COMPANIES AND BOND ISSUERS The domestic securities exchange should mandate publicly listed companies and bond issuers to publish CSR reports on a regular basis to disclose environmental information. Such disclosure should primarily include information about their emissions and the environmental impacts that have been or could be generated by invested projects or business operations, the efforts to reduce such impacts, and the investments made in environmental protection and energy conservation. For specific disclosure standards, reference may be made to the environmental impact disclosure standards in ISO14000, which are commonly used internationally. Environmental impacts must be expressed quantitatively and must not be limited to qualitative descriptions.

VIII. CREATE A GREEN INVESTORS NETWORK AND A FRAME-WORK FOR INVESTOR SOCIAL RESPONSIBILITY Top domestic institutional investors, such as the largest banks, social security funds, insurance companies, and mutual fund companies, should establish a Chinese "green investors network." The network's primary functions should include promoting environmental impact assessment during the investment decision-making process, taking into consideration the Environmental, Social, and Governance (ESG) principles. These principles can be refined and adapted to suit Chinese investors. The network can also supervise and urge their portfolio companies to assume social responsibility and improve information disclosures, as well as advise the government on policy design to promote green finance and investment.

IX. INCREASE CONSUMER PREFERENCE FOR GREEN PROD-
UCTS THROUGH PUBLIC EDUCATION Environmental public
education campaigns should be used to raise consumer awareness of
social responsibility, which should also help increase their preference
for green products and their aversion to polluting products. In other
words, a greater number of consumers must be made to prefer green
products even when prices are higher than for nongreen products.

Some of these efforts could include environmental education for
children, disclosure of products' environmental information, establishing role models of environmental protection, encouraging NGOs to
champion environmental awareness, and leveraging public opinion to
censure environmentally unfriendly consumer behavior, among others.
In addition, the Chinese media should publicize green projects and
products to help raise awareness and build demand for these products.

CHAPTER FIVE

The Cleanup and Economic Growth

SOME CHINESE POLICYMAKERS remain concerned about whether ambitious efforts to tackle air pollution will lead to a significant economic slowdown that could trigger unemployment and catalyze social instability. Others worry about how to pay for the pollution cleanup without breaking fiscal coffers. Moreover, if businesses and consumers are required to shoulder these costs in accordance with the "user pays" principle, will that create inflationary pressure?

If tackling pollution mainly takes the form of administrative policy— for example, forcing closure of factories and construction sites and establishing mandatory limits on automobile traffic—this will certainly cause a substantial blow to the real economy and may even lead to severe unemployment. The consequences of administrative measures are often only to break down the old without establishing something new, which naturally leads to economic decline. This is because it takes time for resource (capital, labor, and land) allocation to adjust from polluting industries to green industries.

Leveraging economic incentives to gradually reduce the profitability of polluting industries and increase the profitability of green industries is a better way to redirect resources toward green industries, as structural adjustments aimed at cutting PM2.5 pollution can be achieved without major disruptions to the economy as a whole. Many of the economic incentives and policies needed to achieve these goals were examined in earlier chapters. But a key question remains: to what extent will these actions and policies affect the macroeconomy as well as the structure of the economy?

To address this question, we developed a CGE model to analyze the economic impact of our proposed package of actions and policies—in particular, its effects on GDP growth, fiscal balance, and consumer price index (CPI). To reiterate, these policies and actions include raising the cost of industrial land, lowering the indirect tax burden on the services industry, hiking pollution fees, increasing the resource tax on coal, instituting a carbon tax, increasing subsidies for clean energy, raising automotive fuel quality and emissions standards, implementing license plate auction and congestion fees, and substantially ramping up investment in rail and subways.

We conclude that the impact of these actions on China's GDP growth is very modest. In terms of the fiscal impact, subsidies for clean energy will be partially offset by higher resource taxes and environmental taxes and fees, and will not significantly increase the fiscal deficit. Although our recommended policies will push the producer price index (PPI) modestly higher, their effect on the CPI will be very small. This is because the transmission from PPI to CPI is limited, and the reduction in the tax burden on the services sector will also partially offset upward pressure on consumer prices.

In terms of the impacts on industry, we conclude that the traditional coal industry will experience the greatest negative effects, while profits and growth will also be squeezed in many energy-intensive heavy industries. Meanwhile, growth in industries such as natural gas and wind power will exceed current expectations. Finally, because of the potential for Chinese auto exports, growth in foreign markets will partially offset the effects of lower domestic demand for automobiles.

I. Estimating the Effects of Pollution Reduction on the Economy

1. Overview of the CGE Model

CGE models, which were built based on Walrasian equilibrium and incorporating input-output tables, are generally used for policy analysis. Such models typically use a large set of simultaneous equations to describe the relationships of supply, demand, and the market. In the past decade or so, we have used various CGE models to assess the impacts of trade policy, exchange rate policy, pension reform, fiscal policy, and external shocks on the Chinese economy.

In this study, we extended the ORNAIG model, a static CGE model developed at Monash University in Australia, used 2007 input-output data in China, and incorporated other Chinese macroeconomic and industry data to build a static CGE model with 135 sectors in China. The model can be used to simulate the economic effects of various policy changes, including the effects on variables such as national output, industrial output, employment, and prices (see more details in the appendix).

2. Policy Assumptions in the CGE Model Analysis

To simulate the effects of the proposed emissions reduction actions (primarily structural adjustment policies) on the economy, we simplified many policy recommendations so that they could be realized in the model (see figure 5.1). Our policy change (shock) assumptions are as follows:

1. The price of industrial land in China would double due to measures that reduce the supply of industrial land. In the model, this is manifest as an increase in the cost to industrial firms of investing in factory buildings, and a corresponding reduction in the price of residential and commercial land due to measures that boost supply.

2. The indirect tax rate for the services sector would fall by 1 percentage point.

3. Power companies and industrial firms would be encouraged to install end-of-pipe control equipment as a result of higher fees on SO_2 and NOx emissions and stronger enforcement of desulfurization and denitration. In our simulation, this is achieved by increasing capital investment and capital depreciation of coal producers (see figure 5.1). Because of increased adoption and operation of such equipment by power generation and coal companies, the capital input costs for these firms would increase by 5 percent.

4. The resource tax rate on coal would be increased from 0.7 percent to 5 percent and also increased by 3 percentage points for commodities such as petroleum and metals. Resource taxes are

incorporated into the model at the output price level, causing the price of inputs such as coal to increase.

5. A carbon tax would be imposed, with ad valorem rates of 1.8 percent for coal and 0.4 percent for oil and natural gas, which are calculated based on the intensity of emissions and current energy prices.

6. The additional revenue raised by increasing the coal resource tax (i.e. the incremental portion) would be used to bolster subsidies for clean energy, the equivalent of instituting a negative indirect tax on the clean energy sector. According to our forecasts, annual coal consumption will be approximately 4 billion tons over the next several years. At a price of 600 yuan ($97)/ ton and an increase in the coal resource tax rate by 4.3 percentage points, the incremental resource tax collected on coal will be about 100 billion yuan ($16.1 billion). This additional tax revenue will be used to subsidize new energy, which in the model includes wind, solar, and hydropower but does not include natural gas. In the model, subsidizing natural gas is achieved by lowering the indirect tax burden on new energy industries, so that clean energy sources become substitutes for coal (see figure 5.1).

7. The quality of automotive fuel products would be improved. In the model, the effect on the economy is to raise the price of fuel by increasing the input costs.

8. Automobile sales would be curtailed through measures such as increasing fuel emissions standards, introducing license plate auctions, and implementing congestion fees. In the model, the effect on the economy is achieved by raising direct taxes on automobiles, thereby increasing the cost of automobile consumption.

9. Investment in public transportation, such as subways, would increase. Sources of financing include allocating a portion of the revenue from license plate auctions and congestion fees. It is assumed that on a national basis, this additional annual revenue will total 30 billion yuan ($4.8 billion).

10. The effect of these changes in fiscal policy is budget neutral— that is, total increase in revenue equals the total increase in expenditure.

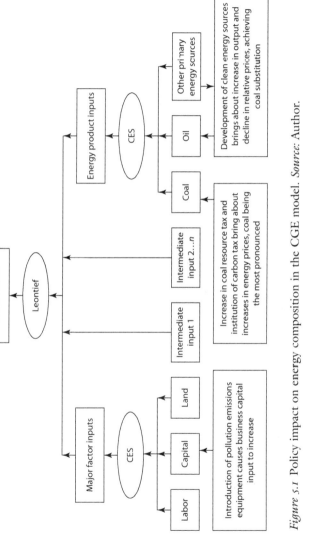

Figure 5.1 Policy impact on energy composition in the CGE model. *Source:* Author.

3. Impact on Economic Growth

The effect of structural adjustment policies on overall economic growth is very modest, according to results from our CGE model analysis (see box).

CGE Model Simulation Results

The simulations of the nine policies ("shocks") described above were carried out to obtain results comprising a series of effects at the macroeconomic level and at the industry level. All of these results are percentage changes in the economic variables resulting from the policy "shocks" compared to the baseline scenario (i.e., the scenario in which policies remain unchanged) during the period of policy implementation.

Viewed from the perspective of economic growth, the policy shocks will cause real GDP to decline a cumulative 0.12 percent, primarily due to a decline in total factor productivity (TFP). For example, because environmental protection requirements are added to production processes, greater input is required for the same output. However, if the implementation phase for these policies spans three to five years, then the effect on annual GDP growth is negligible.

In terms of prices, the policy package above will drive up PPI approximately 0.6 percent, but will simultaneously lower the CPI by approximately 0.3 percent. The reason for the divergence lies in the fact that the policy package will raise the price of commodities and other inputs directly associated with the PPI, such as coal, industrial products, machinery, and equipment. But at the same time, lower indirect tax burden on the services sector, changes in land supply policy, and the slight deceleration in economic growth will all put downward pressure on CPI.

Based on the expenditure approach to calculating GDP, due to the effects of factors such as decline in the CPI (positive), lower services sector tax burden (positive), and deceleration in GDP growth (negative), real household consumption will increase slightly. On the other hand, due to the effects of positive factors including increased investment in environmental protection equipment, new energy industries, and urban public transportation infrastructure, real investment will continue to increase despite the decline in industrial output and slower GDP growth. Affected by rising PPI,

---

TABLE 5.1
Cumulative Change in Economic Variables (%)

CPI	–0.26
PPI	0.55
Real GDP	–0.12
Real consumption	0.13
Real investment	0.25
Real exports	–0.18
Real imports	0.78

Source: CGE model simulation.

export costs will go up, resulting in a decline in actual volume and a corresponding increase in imports (see table 5.1).

The policies' impact on the industry structure is greater than their effect on the economy as a whole. From the perspective of industrial output, the real output of coal extraction and processing, as well as thermal power industries, will decline approximately 12 percent compared to the baseline scenario, while clean energy power generation will increase 60 percent, essentially matching the energy structure adjustment outcome we anticipated. At the same time, as a result of these policies, heavy industry and the thermal power industry together as a proportion of GDP will decline by 0.6 percentage points, and the services sector's proportion will increase. This comports with the industry structural adjustment we expect for the coming few years. The industries most negatively affected by these policies include thermal power, coal extraction, automobiles, steel, petrochemicals, and basic chemicals. Meanwhile, output will increase in services industries, including urban public transportation, real estate, travel, software, and household services (see table 5.2).

Our current CGE model is unable to simulate the changes in the urban transportation structure (e.g., the change in subway travel as a proportion of total travel), because the industry classification in the model does not single out subways but simply incorporates it under urban public transport. Also, our assumption in the model that revenue from license plate auctions and congestion fees will be used to support subway construction is only one way in which funding for subway construction can be increased. Other actions that should be encouraged simultaneously include the issuance of municipal bonds and establishment of PPPs to leverage private capital. If these actions are taken together, the proportion of subway travel will grow faster.

TABLE 5.2
Changes in Real Output of Major Industries (%)

Thermal power industry	-12.63	Public administration and social organizations	0.07
Coal extraction and processing	-11.62	Cement, lime, and plaster manufacturing	0.09
Automobile manufacturing	-2.15	Wholesale and retail industries	0.10
Ferrous metal extraction and processing	-2.05	Insurance	0.13
Nonferrous metal extraction and processing	-1.95	Telecom and other information services	0.13
Oil and nuclear fuel processing	-1.69	Food service	0.14
Basic chemical raw materials manufacturing	-0.72	Agriculture	0.14
Rubber products	-0.53	Commercial services	0.24
Steel smelting	-0.49	Construction	0.26
Road transportation	-0.39	Household services	0.39
Electronic components manufacturing	-0.20	Travel	0.47
Paper and paper products	-0.06	Real estate	0.58
Air transportation	-0.04	Urban public transportation	3.33
Financial services	0.01	Clean energy power generation	59.64

Source: CGE model simulation.

International experience also supports our view that emissions reduction does not necessarily lead to a substantial economic slowdown. The United Kingdom is a case in point. From the 1950s to the 1970s, coal consumption in the United Kingdom fell by roughly 40 percent and air pollution declined by nearly 80 percent, but GDP growth remained essentially stable (see figure 5.2). Germany, too, was able to dramatically reduce pollution without severely disrupting economic growth.

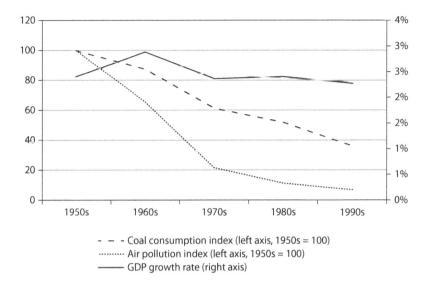

Figure 5.2 GDP growth, coal consumption, and air quality in the United Kingdom. *Sources:* Author's estimates; UK Office of National Statistics; UK Department of Energy and Climate Change; AEA Environmental Protection.

In the two decades from 1990 to 2010, Germany's total coal consumption fell by 40 percent, while SO_2 and NOx emissions plummeted by 90 percent and 55 percent, respectively. At the same time, average GDP growth remained virtually the same as it had been for the preceding twenty-year period (1970 to 1990). Moreover, during the early 1990s, a period in which Germany intensified pollution mitigation efforts, the country was able to manage 4 percent GDP growth.

The key reason that tackling air pollution will not necessarily come at the significant expense of economic growth is because the various emissions reduction measures are primarily structural in nature. In other words, the actions hurt growth in some industries but stimulate growth in others, so the effects largely offset each other, thus limiting the overall impact on the economy.

To illustrate, figure 5.3 shows estimated changes in output in certain industries affected by the proposed policy measures. Based on the CGE model simulation, compared to the baseline scenario of no policy changes, total output for the coal and automobile industries will decline by around 230 billion yuan ($36 billion), measured in 2011 constant

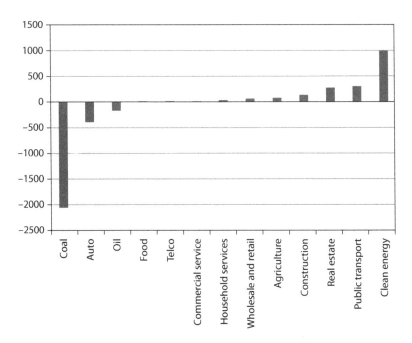

Figure 5.3 Impact of recommended policies on sector output for 2020 (in 2011 constant prices). *Source:* Author's estimates.

prices. But at the same time, output of the clean energy, rail, and subway industries will increase by about 200 billion yuan ($32 billion).

4. Impact on Fiscal Revenue and Expenditure

Our analysis indicates that the government will not need to foot the bill for some of the policy measures, because businesses and consumers will bear the costs. Bolstering clean energy will indeed require more extensive central government subsidies, but in the course of using tax policy to rein in pollution, the government will also receive new sources of fiscal revenue. The additional expenditure and additional revenue will offset each other, making the structural adjustment efforts basically revenue neutral.

Currently, China's clean energy subsidies primarily include a price subsidy for wind power of 0.21–0.28 yuan ($0.03–$0.045)/kWh, a price subsidy for solar power of 0.5 yuan ($0.08)/kWh, average subsidy of about 100,000 yuan ($16,129) per EV (with additional subsidies for the construction of charging stations), and a subsidy of 0.4 yuan ($0.06)/m³ for shale gas extraction. Assuming these subsidy rates remain unchanged, and that an additional subsidy is provided for natural gas-generated electricity, then based on our latest growth forecast for different types of clean energy, fiscal subsidies for clean energy as a proportion of total fiscal expenditure will increase from around 0.5 percent in 2012, or 60 billion yuan ($9.7 billion), to 1.2 percent in 2015, or 185 billion yuan ($30 billion) (see table 5.3).

At the same time, proposed policies to raise the coal resource tax and pollution levies and to implement license plate auctions will all boost fiscal coffers. If the coal resource tax is increased to 5 percent, SO_2 and NOx emissions charges are doubled, and license plate auction systems are put in place in the next ten largest cities after Shanghai, Guangzhou, Tianjin, and Hangzhou—which all have auction systems now—we estimate that new revenue as a proportion of total fiscal revenue will increase from 0.4 percent in 2012, or 43 billion yuan ($6.9 billion), to 1.2 percent in 2015, or 173 billion yuan ($28 billion) (see table 5.4).

5. Impact on Inflation

Costs for businesses will certainly rise if the recommended actions and policies are enforced. The cost pressures will be predominantly felt in the coal, power, and other heavy industries, as well as in the auto sector.

While consumers would also feel the impact of the rising cost of cars because of new taxes and a license auction system, consumers will pay less for services due to lower taxes on the services sector. In addition, increasing the supply of residential and commercial land will help to lower CPI. More specifically, the impacts on consumers are as follows:

- Electricity that includes desulfurization has an additional price premium of 0.015 yuan ($0.002)/kWh on top of the 2013 base tariff (without the desulfurization) of about 0.4 yuan ($0.065)/kWh in most provinces. Electricity after denitration has an additional tariff premium of 0.01 yuan ($0.002)/kWh, according to the State Electricity Regulatory Commission.
- Upgrading to the National V fuel standard will raise gasoline prices by 0.34 yuan ($0.055)/liter. Based on data from the ICIS energy network, 64 percent of Sinopec's and 23 percent of CNPC's refining capacity can now meet the National IV standard. The two major oil companies will need to invest another 50–60 billion yuan ($8.1–$9.7 billion) to fully upgrade all refining capacity to

TABLE 5.3
Government Fiscal Subsidies for Clean Energy

Energy source	Subsidy	2012 output	Total subsidy in 2012	2015 output forecast	2015 subsidy forecast
Wind power	0.21–0.28 yuan/kWh	100 billion kWh	25 billion yuan	190 billion kWh	47.5 billion yuan
Solar power	0.5 yuan/kWh	9 billion kWh	4.5 billion yuan	25 billion kWh	12.5 billion yuan
EVs	100,000 yuan/vehicle	120,000 vehicles	1.2 billion yuan	500,000 vehicles	50 billion yuan
Shale gas extraction	0.4 yuan/m³	30 million m³	12 million yuan	6.5 billion m³	2.6 billion yuan
IGCC/natural gas power generation	IGCC power generation 0.2 yuan/kWh;* natural gas 0.35 yuan/kWh	Natural gas power generation about 90 billion kWh†	About 30 billion yuan	IGCC power generation 60 billion kWh/natural gas power generation 180 billion kWh, together accounting for 5% of power generated	75 billion yuan
Total			**60 billion yuan**		**185 billion yuan**
Proportion of total fiscal spending			**0.48%**		**1.19%**

*Amount recommended in this study.

† Total power generated in 2012 was 4.9377 trillion kWh, approximately 1.8 percent of which was power generated from natural gas (World Bank data for 2011).

Notes: IGCC = integrated gasification combined cycle.

Source: China Electricity Council.

TABLE 5.4
New Sources of Fiscal Revenue

Policy	Action	2012 revenue	2015 revenue forecast
Coal resource tax	Increase by 4.3 percentage points (from 0.7% to 5%)	4 billion tons of coal consumed each year, price of 600 yuan/ton; calculated at 0.7%, total revenue of 17 billion yuan	Additional 100 billion yuan for a total of 117 billion yuan.
Pollution fees	Double pollution fees for SO_2 and NOx emissions	Total pollution fees nationwide were 20 billion yuan in 2011	Initially increase by 20 billion yuan to 40 billion yuan; thereafter, as businesses increase investment in pollution equipment and reduce emissions, additional revenue from pollution fees will fall. Over the long term, it can be considered stable at 20 billion yuan.
Automobile license plate auctions	Implementation of automobile license plate auctions in nearly ten major cities	Annual average revenue from automobile license plate auctions in Shanghai was 6 billion yuan in 2011 and 2012	Estimated new automobile license plate auction revenue of approximately 30 billion yuan, for a total of 36 billion yuan.
Total		43 billion yuan	173 billion yuan
Proportion of total fiscal revenue		0.37%	1.16%

Source: Author's estimates.

meet National IV. To comply with the stricter National V standard, Sinopec and CNPC will need to invest an additional 88.7 billion yuan ($14.3 billion) and 129.6 billion yuan ($21 billion) for upgrades, respectively. This means that the two oil giants will each need to invest 22.2 billion yuan ($3.6 billion) and 32.5 billion yuan ($5.2 billion) a year from 2013 to 2017 to upgrade refineries, thereby raising gasoline costs.

- Moving to National V emissions standards will increase the cost of manufacturing a standard 2.0-liter air discharge capacity vehicle by approximately 2,000 yuan ($322.58), according to the standards contained in the draft regulation "Light Automobile Pollutant Emissions Limits and Measurement Methods (China Stage 5)."
- Lowering the indirect tax rate on the services sector by 1 percentage point will lead to a reduction of about 0.5 percentage points in consumer prices for services.

Based on the data above, we assumed that the enforcement of the National V fuel standard would lead to a 5 percent increase in gasoline prices, the enforcement of the National V emissions standard would lead to a 3 percent increase in automobile prices, the increase in the coal resource tax and desulfurization installation and operation would lead to a 5 percent hike in electricity prices, and consumer prices for services would drop 0.5 percent.

We then used the CGE model to simulate the effects of these price increases on the PPI and the CPI. The results showed PPI up a cumulative 0.55 percent and CPI down a cumulative 0.26 percent. If the policy implementation period is five years, then the annual impact on prices is limited.

6. Impact on Major Industries

1. COAL In our view, China's conventional coal consumption will decline by a cumulative 14 percent from 2013 to 2030 and output will fall dramatically as well. If this projection is realized, China could well transform itself from a net importer of 200 million tons of coal in 2013 to once again becoming an exporter of coal. This will put downward pressure on global coal prices, since China's coal imports account for one-quarter of global seaborne coal trade.[2] But if coal gasification can partially sustain

the domestic demand for conventional coal, then the shock to the coal industry could be somewhat absorbed.

II. HEAVY INDUSTRY Since our proposed policies will drive up prices for electricity, coal, natural gas, and automobiles, certain energy-intensive industries, such as steel, cement, nonferrous metals, and chemicals, will face intensifying cost pressures and falling profit margins (see figure 5.4). Our estimates of profits and losses are based on assumptions of a 5 percent increase in coal prices, a 5 percent hike in electricity prices, a 30 percent increase in natural gas prices, and a 3 percent increase in automobile costs. In general, small- and medium-sized businesses in these industries will face greater pressure, because larger businesses have more or less already conformed to stricter environmental standards. This also means that industry consolidation will accelerate as smaller firms seek to exit the market.

III. CLEAN ENERGY We have already established that natural gas consumption growth needs to be maintained at 14 percent a year from 2013 to 2020, faster than the 10 percent annual growth we previously projected. Producers will of course be the chief beneficiaries of more robust natural gas consumption, including the three national oil companies, as well as a number of pipeline operators and distributors.

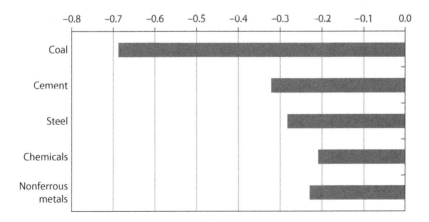

Figure 5.4 Changes in pretax profits of heavy industries under structural adjustment. *Source:* CGE model.

Wind power should also expand rapidly, maintaining 20 percent average growth from 2013 to 2020. In addition, we recommend relocating a portion of thermal power plants to sparsely populated areas in central and western China, as well as increasing the construction of ultra high-voltage transmission lines because China's wind and hydro-power resources are concentrated in the relatively remote southwest and northwest, while demand is concentrated along the coast. Now that the wind power manufacturing industry has emerged from a cyclical trough, the leading companies in this industry will certainly benefit.

IV. RAIL TRANSPORTATION Under the structural adjustment scenario, total railway length will increase by 60 percent and total subway length by 400 percent between 2013 and 2020, which are 15 percent and 40 percent higher, respectively, than are called for in existing government plans. If these targets are realized, average annual investment in rail construction will increase from the currently planned 530 billion yuan ($85.5 billion) to 770 billion yuan ($124.2 billion) from 2013 to 2020, while subway construction growth will increase from 3 percent to 17 percent a year over the same period.

V. CLEAN COAL TECHNOLOGIES Three industries will benefit from the wider adoption of desulfurization and denitration technologies and adherence to more stringent emissions standards. The first is the environmental pollution monitoring and control industry. We estimate that in order to gradually expand PM2.5 monitoring to all cities, China will need to invest 2 billion yuan ($323 million) to acquire equipment and technology. The second is the central heating industry. For example, Beijing has plans to scrap 1,600 coal-fired boilers and 44,000 heating systems in older, single-story houses and replace them with central heating systems. The third is the desulfurization and denitration equipment industry, which is estimated to see annual investment of at least 60 billion yuan ($9.7 billion) in the next few years.

VI. AUTOMOBILE INDUSTRY While many of the policy recommendations will negatively affect car sales, China can aim to become a major car exporter over the next ten to twenty years due to technological progress and improved competitiveness. At the same time, the rapid development of EVs and public transit buses will benefit parts of the automobile industry.

For instance, current Chinese passenger vehicle exports account for only 3 percent of output. Auto analysts estimate that by 2030, this figure could rise to 20 percent, at which time Chinese auto manufacturers could control 6–7 percent of global market share. (In contrast, major Japanese carmakers export more than half of their total output.) Assuming this market share can be achieved, Chinese car exports will reach 6 million units by 2030. This means that although growth in domestic sales will slow, annual output growth could be maintained at about 7 percent, an impressive rate compared to 2–3 percent growth for major Western automakers.

In terms of EVs, we estimate that the current industry forecast of 1 million EVs by 2020 can be multiplied by a factor of three or more when technological improvements and government support become more robust, which will certainly benefit EV manufacturers. In the near term, however, the main obstacle is the lack of EV charging infrastructure.

APPENDIX

The CGE the Model

1. Overview of CGE model

The computable general equilibrium (CGE) model used in this study is a policy analysis model based on Walrasian equilibrium and incorporates an input-output table. A CGE model typically consists of a large set of equations to describe the relationships between supply, demand, and market equilibrium. Optimization conditions, such as producer profit maximization and consumer utility maximization, are used to solve the equations to obtain the set of quantities (e.g., commodity and primary factor quantities) and prices (e.g., commodity prices, capital returns, and wages) at which equilibria are achieved in all markets.

In 1960, Leif Johansen built the first CGE model. His model was based on input and output data for Norway, incorporating twenty cost-minimizing industry sectors and one utility maximizing household sector. Following Johansen's contribution, there was a surprisingly long hiatus in the development of CGE modeling with no further significant progress until the 1970s. After a long period during which leading economists developed and refined theoretical propositions of general equilibrium, Arrow and Hahn (1971) more comprehensively summarized the existence, uniqueness, optimality, and stability of the Walrasian equilibrium solutions.

In the 1970s, Western countries experienced a series of major shocks, including a dramatic rise in energy prices, the collapse of the Bretton Woods system, and rapid growth in real wage levels. Traditional econometric models were unable to effectively simulate and evaluate these shocks, but CGE models were able to infer the potential effects of these shocks in the absence of historical experience. Against this backdrop, and in conjunction with improvements in computer programming applications and the availability of economic data, great progress was achieved with CGE models during this period.

CGE modeling has been applied to analyze various economic questions and policies, including fiscal and tax policy, energy and environment, economic growth, and international trade. Research in the energy and environment fields has included studies on the exploitation of primary energy, CO_2 emissions (the greenhouse effect), energy prices, pollution, and development of renewable resources (Dufournaud [1988], Jorgenson [1994], Perroni [1997], Tsigas [1997], Jean [2002], Lin [2010], Beckman [2011], Wianwiwat [2012], and Arshad [2014]).

In this study, we adopted the basic system of equations in the ORNAIG model, a static CGE model developed at Monash University in Australia, and used input-output databases in China for the year 2007 as well as other Chinese macroeconomic and industry data to build a 135-sector static Chinese computable general equilibrium model (DBCGE). It has a database consisting of a large volume of economic data and parameters, as well as a system of equations encompassing more than 10,000 equations.

2. Basic Equation Set for the Static CGE Model

The basic equation set for this model can be divided into six modules: a production module, a domestic demand module, a margin demand module, an import-export trade module, a price module, and a market-clearing module.

I. Production Module

The production module contains data about production input and output. Figure A.1 depicts the composition of the production module, the lower half of which is the production input structure consisting of two levels of nesting used for production inputs. The lower level includes CES composite functions among three major factors and CES functions for the import goods and the domestic goods of intermediate production inputs.

The specific equations are as follows:

$$X_{ij}^{(1)} = CES_{s=\text{dom,imp}} \left\{ \frac{X_{(is)j}^{(1)}}{A_{(is)j}^{(1)}}; \rho_{(ij)}^{(1)}, \beta_{(is)j}^{(1)} \right\} i = 1, \ldots, n \ \ j = 1, \ldots, h \quad (1)$$

$$X_{n+1,j}^{(1)} = CES_{s=\text{lab,cap,lnd}} \left\{ \frac{X_{n+1,s,j}^{(1)}}{A_{n+1,s,j}^{(1)}}; \rho_{n+1,j}^{(1)}, \beta_{n+1,s,j}^{(1)} \right\} i = 1, \ldots, n \ \ j = 1, \ldots, \quad (2)$$

Equation 1 states that input $X_{ij}^{(1)}$ is the CES composite of the corresponding domestic commodity (s=dom) and import commodity (s=import) components. Equation 2 states that the basic factor input is the CES composite of labor (s=lab), capital (s=cap), and land (s=land).

The upper level of the production inputs portion comprises Leontief sets for major factor inputs, other consumption, and intermediate product inputs. The specific equation is as follows:

$$\text{Leontief}_{i=1,2,\ldots,n+2} \left\{ \frac{X_{ij}^{(1)}}{A_{ij}^{(1)}} \right\} = A_j^{(1)} Z_j \ i = 1, \ldots, n \ \ j = 1, \ldots, h \quad (3)$$

Equation 3 states that production requires $n + 2$ types of inputs, of which the first n input types are intermediate input commodities, the $n + 1$ input type is primary factors, and the $n + 2$ input type is other costs. Optimization of producer behavior is reflected in the model as minimization of production costs at the production input level:

Min:

$$\sum_{i=1}^{n} \sum_{s=1}^{2} P_{(is)j}^{(1)} \times X_{(is)j}^{(1)} + \sum_{s=1}^{3} P_{n+1,s,j}^{(1)} \times X_{n+1,s,j}^{(1)} \quad (4)$$

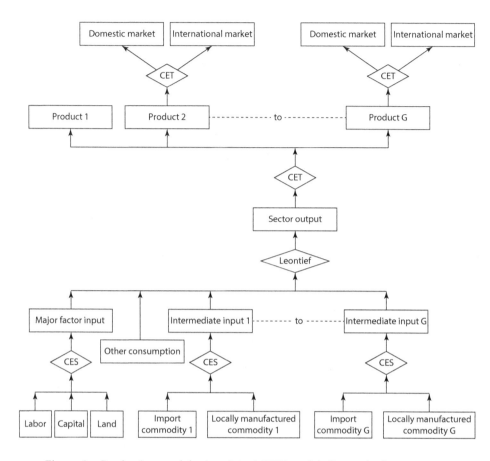

Figure A.1 Production modules in original CGE model. *Source:* Author.

The upper portion of figure A.1 is the production output structure. As can be seen in the figure, there are two levels of CET nesting in the production output portion. The first level represents decisions made by producers regarding the combination of output products based on the relative prices of various input products. The equation is as follows:

$$CET_{s=1,2,\ldots,n}\left\{\frac{X_i}{A_i}; \rho_i, \beta_i\right\} \qquad (5)$$

The second level represents decisions made by producers regarding changes in the sale of their products in international and domestic markets, based on the relative changes in prices on international and domestic commodities markets. The equation is as follows:

$$CET_{s=1,2} \left\{ \frac{X_i^{(s)}}{A_i^{(s)}}; \rho_i, \beta_i^{(s)} \right\} \quad i = 1, \ldots, n \tag{6}$$

In the production output portion, optimization of producer behavior is reflected as revenue maximization—that is, the output decisions provided by equation 5 and equation 6 must satisfy the following:

$$\text{MAX:} \sum_{i=1}^{n} \sum_{s=1}^{2} P_i^{(s)} \times X_i^{(s)} \tag{7}$$

II. Domestic Demand Module

The domestic demand module comprises four categories of demand: investment demand, household consumption demand, government spending demand, and inventory demand.

Similar to the production decisions, the objective of investment decisions is to minimize costs, that is:

$$\text{Min:} \sum_{i=s}^{n} \sum_{s=1}^{2} P_{(is)j}^{(2)} \times X_{(is)j}^{(2)} \tag{8}$$

Under the budget constraint (equation 9), households pursue maximization of utility (equation 10). Utility is expressed using the Klein-Rubin function:

$$\text{s. t.} \sum_{i=1}^{n} P_i^{(3)} \times X_i^{(3)} = EP \tag{9}$$

$$\text{MAX:} \ U = \sum_{i=1}^{n} d_i \times \ln\left(\frac{X_i^{(3)}}{A_i^{(3)}} - \theta_i \right) \tag{10}$$

Government spending is expressed using three types of equations: government spending-exogenous, government deficits-exogenous, and government spending linked with household consumption. Inventory demand is generally linked to GDP.

III. Margin Demand Module

The model assumes that margin demand is determined by commodity circulation volume and margin demand coefficients. Equation 11 is for the margin demand decision.

$$X_{(r1)}^{(is)^{jk}} = A_{(r1)}^{(is)^{jk}} X_{(is)j}^{(k)} \quad i, r = 1, \ldots, g \; j = 1, \ldots, h \; k, s = 1, 2 \qquad (11)$$

IV. Import-Export Trade Modules

In the import module, the model uses the Armington assumption to build the CES composite of domestic products and import products. At the same time, the model determines the price of import products exogenously. At this price, there is an unlimited supply of import products.

In the export module, the model uses export demand that is described using a downward sloping curve of fixed price elasticity.

$$X_i^{(4)} = FQ_i^4 \left[P_i^4 / \left(\Pi \times FP_i^4 \right) \right]^{\sigma_i^4} \quad i = 1, \ldots, n \qquad (12)$$

V. Price Module

The pricing system of this model is based on the following two assumptions. First, the market exhibits competitiveness (i.e., production and sales activities are all zero profit). Second, the producer price for each type of commodity is unique, does not vary in different commodity production sectors or users (for domestically produced products), and does not vary for different importers (for imported products). In this way, the assumption in the model that zero-profit conditions and constant returns to scale for production activities remain unchanged means that the producer price is only the input price function. The assumption in the model concerning the zero-profit condition means that the buyer price is the sum of the producer price of the commodity, indirect tax, and margin demand costs, specified as follows:

Commodity production price equation:

$$p_j - a_j = \sum_{i=1}^{n} H_{ij}(p_{ij} + a_{ij}) + H_j^F p_j^F + H_i^0 p_i^0 \tag{13}$$

Composite commodity price equation:

$$p_i^{(n)} = \sum_{s=1}^{2} S_{is}^{(n)}(p_{is}^{(n)} + a_{is}^{(n)}) \tag{14}$$

Capital price equation:

$$p_j^{(2)} - a_j^{(2)} = \sum_{i=1}^{25} H_{ij}^{(2)}(p_{ij}^{(2)} + a_{ij}^{(2)}) \tag{15}$$

Labor price equation:

$$p_{n+1,1m}^{(1)} - a_{n+1,1,m}^{(1)} = \sum_{j=1}^{25} H_{n+1,1,m}^{(1)}(p_{(n+1,1,m)j}^{(1)} + a_{(n+1,1,m)j}^{(1)}) \tag{16}$$

Import price equation:

$$p_i^{(1)} = pw_i^{(1)} + \pi + t_i^{(1)} \tag{17}$$

Export price equation:

$$p_i^{(4)} = pw_i^{(4)} + \pi + t_i^{(4)} \tag{18}$$

Price equation for margin and indirect tax:

$$P_i^{Purchase} X_i = P_i^0 X_i + P_i^0 X_i T_i + P_i^{Margin} X_i \tag{19}$$

VI. Market-Clearing Module

General equilibrium models require the commodity market and the factor market to be in equilibrium simultaneously. That is, they are in the market-clearing state, in which:

1. The model assumes that the commodity market must be cleared— that is, the supply and demand of domestically produced products and imported products must be equal. Equation 20 presents the market-clearing condition for domestically produced products. The left side of the equation represents aggregate supply of the domestically produced product market, and the right side of the equation represents market demand for domestically produced products, comprising intermediate input demand, investment demand, household consumption demand, export demand, government spending demand, and inventory demand.

Equation 21 presents the market-clearing condition for imported products. The left side of the equation represents aggregate supply of the imported product market, and the right side of the equation represents market demand for imported products, comprising intermediate input demand, investment demand, household consumption demand, and government spending demand.

$$
\begin{aligned}
x_i^{(D)} = \sum_i & \left(S_i^{(1D)} x_i^{1D} + S_i^{(2D)} x_i^{2D} \right) + S_i^{(3D)} x_i^{3D} + S_i^{(4D)} x_i^{4D} \\
& + S_i^{(5D)} x_i^{5D} + 100 \times PO_i^{(6D)} \times delx_i^{(6D)} / V_i^{(6D)}
\end{aligned}
\tag{20}
$$

$$
x_i^{(1)} = \sum_j \left(S_i^{(1I)} x_i^{(1I)} + S_i^{(2I)} x_i^{(2I)} \right) + S_i^{(3I)} x_i^{(3I)} + S_i^{(5D)} x_i^{(5D)}
\tag{21}
$$

2. Factor market-clearing is reflected differently in the labor market and the capital market. Labor is highly mobile among various industries. Labor market-clearing requires the sum total of labor inputs in the various industries to equal the aggregate supply of labor, shown in equation 22:

$$
\sum_j x_{(n+1,1,m)j}^{(1)} = L_m^s
\tag{22}
$$

Capital mobility from industry to industry is poor, therefore capital market-clearing requires that the capital input in each of the industries equal the supply of capital in each respective industry. This is reflected in equation 23:

$$x^{(1)}_{(n+1,2)j} = K^s_j \quad j = 1,\dots,h \tag{23}$$

3. Treatment of Energy Substitution

In the original model as depicted in figure A.1, each type of energy products and their corresponding inputs are inputs into production in the form of Leontief sets, and there is no substitution relationship. However, in this study, in order to simulate the effects of various types of government policies on the energy structure and consumption, it was necessary to assume that the different types of energy can have substitutes (e.g., coal can be replaced by oil, natural gas, and other clean energy sources). Thus, for this study, the production structure depicted in figure A.1 was converted to the production structure depicted in figure A.2.

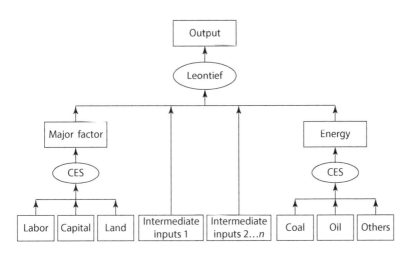

Figure A.2 Production modules in CGE model integrating green fiscal policies.
Source: Author.

In order to reflect substitution, we formed a composite of the various energy inputs using the CES equation format to obtain equation 24:

$$X_{ej}^{(1)} = CES_{s=\text{oil,coal,others}} \left\{ \frac{X_{(es)j}^{(1)}}{A_{(es)j}^{(1)}}; \rho_{(ej)}^{(1)}, \beta_{(es)j}^{(1)} \right\} \quad j = 1, \ldots, h \quad (24)$$

The specific process used to implement this in the GENPACK software consisted of the following steps:

1. Define the set of energy commodities.

Set

COM read elements from file BASEDATA header *"COM"*;

The statement above defines the set of all commodities and assigns its address in the database.

Set

EOM read elements from file BASEDATA header *"ECOM"*;

The statement above defines the set of energy commodities and assigns its address in the database.

Subset EOM is subset of COM;

The statement above defines the set of energy commodities as a subset of the set of all commodities.

2. Define the substitution elasticity among energy commodities and the summation parameters for all energy commodities.

Coefficient

(all,c,EOM)SIGMAE(c);

Read SIGMAE from file BASEDATA header *"SIGE"*;

The statements above define the substitution elasticity of energy commodities and assign its address in the database.

Coefficient

(all,c,EOM)SE(c);

FORMULA

(all,c,EOM)SE(c)= MAKE_I(c)/sum{e,EOM,MAKE_I(e)};

The statements above define the summation parameters for energy commodities and perform a summation of the total volume of energy commodities.

3. Define the weighted average price of energy commodities.

Equation E_p1_se

(all,i,IND)p1_se(i) = sum{c,EOM, SE(c)*[p1_s(c,i)]};

4. Convert the CES substitution relationship among energy commodities into the form of a linear equation set.

(all,c,EOM)(all,i,IND) x1_s(c,i) − a1se(c,i)

= x1_se(i) - SIGMAE(c)*[p1_s(c,i) + a1se(c,i) − p1_se(i)];

Leontief grouping was then performed on the equation set as a whole, together with the other inputs, to obtain equation 25, which corresponds to equation 3 in the original equation set. This states that production requires $n + 3$ types of input, of which the first n input types are intermediate inputs, the $n + 1$ input type is primary factors, the $n + 2$ input type is other costs, and the $n + 3$ input type is energy commodities.

$$\text{Leontief}_{i=1,2,\ldots,n+3}\left\{\frac{X_{ij}^{(1)}}{A_{ij}^{(1)}}\right\} = A_{j}^{(1)}Z_{j} \quad i = 1, \ldots, n \ \ j = 1, \ldots, h \quad (25)$$

The process for implementing this in the GENPACK software consisted of the following:

1. Define the set of nonenergy commodity inputs.

Set NEOM = COM -EOM;

The statement above defines the set of nonenergy commodities and describes the set of nonenergy commodities as a complementary set of the set of all commodities.

2. Split the original Leontief equation.
The original equation:

Equation E_x1_s # *Demands for commodity composites* #

(all,c,COM)(all,i,IND) x1_s(c,i) – [a1_s(c,i) + a1tot(i)] = x1tot(i);

. . . is split to form the equation for nonenergy commodities:

Equation E_x1_s # *Demands for nonenergy commodity composites* #

(all,c,NEOM)(all,i,IND) x1_s(c,i) – [a1_s(c,i) + a1tot(i)] = x1tot(i);

. . . and the equation for energy commodities:

Equation E_x1_se # *Demands for energy commodity composites* #

(all,i,IND) x1_se(i) – [a1sec(i) + a1tot(i)] = x1tot(i)

Case Studies and Green Finance

CHAPTER SIX

Case Study: Shanghai

Introduction

Shanghai should strive to reduce PM2.5 emissions from roughly 60 μg/m³ in 2013 to 25 μg/m³ by 2027, meeting the WHO interim target-2 standard.[1] Our findings show that of the 35 μg/m³ emissions reduction required for Shanghai to reach its target, environmental actions and emissions reductions in peripheral areas will contribute to a reduction of 18 μg/m³, while structural adjustments are needed to achieve the remaining reduction of 17 μg/m³.

More specifically, our recommendations for structural adjustments are as follows:

- move high-polluting steel and petrochemical production out of the city;
- slow the annual growth in the vehicle fleet size to 2 percent;
- limit the term of validity for license plates and/or introduce a congestion fee;
- increase total subway rail length to 1,200 km;
- increase the number of subway cars and boost carrying efficiency;
- attract private capital and issue municipal bonds to fund public transportation;
- raise taxes to increase the cost of conventional coal;
- substantially increase subsidies for clean energy and explore the use of a trading system for new energy vouchers;

- improve the structure of marine transport cargo and increase the proportion of transshipment cargo;
- require ships arriving at port to use clean energy and establish emissions control zones;
- promote the use of onshore power by ships in port and replace oil with natural gas/LNG for mechanical equipment at piers.

Shanghai Should Take the Lead

Since the national average urban PM2.5 level needs to be reduced to 30 µg/m³ by 2030, it is imperative that Shanghai—the most economically developed metropolis in China—stay ahead of the curve in the battle against pollution. In our view, Shanghai is fully equipped to do so, based on three economic rationales.

First, Shanghai's per capita GDP will reach parity with the current level in OECD countries, rising from about $13,000 in 2013 to close to $40,000 in 2027, according to our estimate. The current PM2.5 levels in all OECD countries are below 30 µg/m³, with a mean around 17 µg/m³. Therefore, if Shanghai fails to reduce PM2.5 to 25 µg/m³ by 2027, it will be a laggard in air quality relative to its level of economic development and wealth. For Shanghai, failure to respond to public demands for a healthy city and improved environment would be a notable deficiency in its urban development.

Second, Shanghai will become a predominantly services-oriented economy, which will help reduce PM2.5 substantially. On average, the services sector accounts for about 80 percent of GDP in major developed countries and exceeds 90 percent in many major metropolitan areas around the world. Shanghai, too, will be moving toward that goal over the next couple decades, where conditions will facilitate raising the proportion of services industry to 75 percent of the municipality's GDP. In fact, as of 2012, services value-added already accounted for 60 percent of Shanghai's GDP. Because manufacturing, and particularly heavy industry, is one of the major sources of air pollution in Shanghai, the decline in the proportion of the manufacturing sector and a shift toward services will significantly contribute to Shanghai's PM2.5 reduction.

Third, according to plans at the national level, Shanghai is designated to become an international financial center by 2020—a goal that would be difficult to achieve if PM2.5 levels remain elevated. For instance, the average PM2.5 level in major world financial centers is 20 μg/m³—in New York and London, it is only 13 μg/m³. But Shanghai's current PM2.5 level is 60 μg/m³, a staggering 200 percent higher than the average found in other global financial hubs.

With the expected liberalization of China's capital account and the increasing development of Internet financial services, a prerequisite for Shanghai to become an international financial center is the ability to attract and retain world-class talent, along with their families' willingness to relocate to the city. If Shanghai's PM2.5 remains significantly higher than other global financial centers by 2027, it will be very difficult for Shanghai to compete with other cities for top financial professionals.

International Experience Suggests Shanghai Can Do It Too

In order to demonstrate the feasibility of meeting the 2027 PM2.5 target in Shanghai, the experiences of London, Los Angeles, and Tokyo are instructive. Based on changes in the concentration of major atmospheric pollutants and the reduction in the number of smoggy days per year, it is possible to estimate the number of years it took for these three global cities to achieve notable improvements in air quality (see fig. 6.1).

Based on average annual concentrations of atmospheric smog and SO_2 and the reduction in the number of smoggy days, London's air quality improved markedly between 1956 and 1975, with SO_2 concentrations falling by 70 percent and smog by 80 percent over the nineteen-year period. In addition, based on changes in the number of stage-one pollution alert days and average annual concentrations of major pollutants such as NOx, air quality in Los Angeles improved by about 60 percent from 1977 to 1992.

In Tokyo, SO_2 mitigation efforts adopted from 1968 to the end of the 1970s achieved about 80 percent emissions reductions over twelve years. Moreover, beginning in 1992, programs were implemented to control NOx and suspended particulate matters (SPM). Over the following sixteen years, by 2008, all environmental stations and roadside stations in greater Tokyo had conformed to SPM standards.

Therefore, based on these international experiences in mitigating pollution, we conclude that it is indeed possible for major metropolitan areas such as Shanghai to substantially improve air quality over a period of fifteen to twenty years.

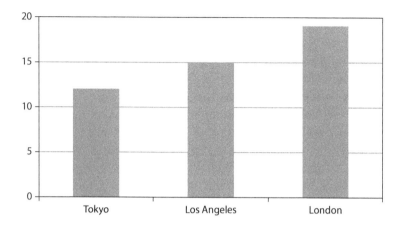

Figure 6.1 Time frames for achieving 60–80 percent reduction in air pollution in major cities (years). *Sources:* Duncan P. H. Laxen and Mark A. Thompson, "Sulfur Dioxide in Greater London, 1931-1985," *Environmental Pollution*, 1987 43(2):103-114; T. Komeiji, K. Aoki, and I. Koyama, "Trends of Air Quality and Atmospheric Deposition in Tokyo," *Atmospheric Environment* Part A, General Topics, 1990 24(8):2099-2103; California Environmental Protection Agency.

I. Adapting the PM2.5 Control Model for Shanghai

We adapted our PM2.5 control model specifically for Shanghai, to analyze the effects of already planned environmental measures and to evaluate whether these measures would enable Shanghai to achieve the proposed PM2.5 emissions reduction target. This adapted model is based on the original PM2.5 control model we developed to quantify the policy effects nationally (see chapters 1–3)—it considered features of the local economic structure and used coefficients based on circumstances specific to Shanghai.

After concluding that environmental measures alone would not be enough to achieve the target, we then performed further simulations that involve a series of proposed economic structural adjustments and policies to meet the target.

Key assumptions of this model are as follows:

- **Target:** Reduce the average level of PM2.5 concentrations from 60 µg/m³ in 2013 to 25 µg/m³ by 2027.

- **GDP growth:** Shanghai's real GDP growth is estimated to gradually slow from 7.5 percent in 2013 to 5 percent in 2027.
- **Coefficients:** Based on historical data, Shanghai's average energy elasticity coefficient is 0.5, its transportation elasticity coefficient is 0.8, and its industry/construction growth elasticity coefficient is 0.9. (In the simulation of the "planned environmental measures only" scenario, we assumed that these elasticity coefficients would not change. However, in the simulation of the "structural adjustment" scenario, we assume that all of these elasticity coefficients will decline.)

We also derived several additional estimates:

1. Based on the assumptions above, without considering changes in the industry and transportation structures, energy consumption growth should average 3 percent and transportation volume growth should average 4.9 percent from 2013 to 2027.
2. Data from the Shanghai Environmental Monitoring Center and relevant academic research show that about 22 percent of Shanghai's primary sources of PM2.5 pollution are due to spillover effects from peripheral provinces and municipalities; around 19 percent come from coal burning and from secondary sulfide and NOx pollution; approximately 31 percent from transportation emissions; 19 percent from noncoal emissions from industry and construction (including VOCs, industrial soot, and construction dust); and the remaining 9 percent from sources such as straw burning, chemical fertilizers, pesticides, tobacco smoking, cooking, forests, and oceans.

Using the assumptions and estimates above as inputs into our PM2.5 control model, we calculated the effects on emissions reduction based first on an improvement in emissions standards and the deployment of clean coal technologies. Second, we estimated the necessary structural adjustments required to achieve the city's pollution reduction target.

Having surveyed Shanghai's planned actions and policies, such as desulfurization, denitration, increasing the quality of petroleum products and automotive emissions standards, and scrapping yellow-label (heavy-polluting) vehicles, among others, we projected that pollution emissions per unit of coal consumed could be cut by 55 percent using clean energy technologies. Moreover, we estimated that average vehicle emissions/km could be slashed by 69 percent via measures such as increasing petroleum product quality, fuel efficiency, and automotive emissions standards, and by routinely replacing catalytic converters.

Even if these end-of-pipe control measures are fully implemented, however, it would still not be possible to meet Shanghai's PM2.5 target by 2027. In order to hit the target, the municipality's economic structure must undergo major changes. That is, Shanghai will need to reduce heavy industry as a proportion of GDP in order to lower the energy elasticity, transportation elasticity, and the construction/industry growth elasticity; alter the energy structure by lowering the proportion of coal and increasing the proportion of clean energy; and dramatically bolster the proportion of rail transport relative to road transport.

While there is obviously more than one way to achieve these structural changes, our combination of paths and policies accounted for factors including heavy industry relocation, the availability of clean energy resources such as natural gas, hydropower, and wind power, technical feasibility, public acceptance (particularly with respect to a more stringent license plate auction system), and international experiences in terms of rail transport density and energy structure.

In the following sections, we use the adapted control model for Shanghai to simulate the effectiveness of planned environmental actions and discuss in detail the effects of the proposed structural adjustment measures.

II. How Effective Are Shanghai's Environmental Measures?

1. Overview of Existing and Planned Environmental Actions

On October 18, 2013, the Shanghai municipal government announced the five-year "Shanghai Clean Air Action Plan (2013–2017),"[2] which, in our view, represented a breakthrough in terms of official thinking on tackling air pollution. Based on this action plan and the previous "Three-Year Action Plan for Environmental Protection and Construction,"[3] Shanghai's existing and planned actions to deal with air pollution can be summarized as follows:

- Reduce coal consumption and achieve negative growth in overall coal consumption by 2017.
- Strictly control total energy consumption, push forward with optimizing energy structure, and accelerate the shift to cleaner energy.
- Accelerate efforts to curb VOC emissions and achieve reduction of 30 percent or more from 2012 levels.
- Speed up the implementation of the National V emissions standard for gasoline-powered light vehicles by the end of 2013 and for diesel and gasoline-powered heavy vehicles by 2015. In addition, phase out yellow-label and older vehicles on an expedited schedule and vigorously promote alternative energy vehicles.
- Extend total rail routes in operation to approximately 600 km by 2014, with the proportion of public transportation commutes exceeding 50 percent in the city center and surpassing 36 percent for the municipality as a whole.
- Optimize port and cargo container transportation systems by increasing the proportion of water-to-water transshipment and water-rail combined shipment of cargo containers. In addition, promote the use of low-sulfur fuel by vessels idling in port and actively encourage the use of onshore power by vessels in port.[4]
- Comprehensively strengthen dust pollution controls in industry and on construction sites, piers, storage yards, and roads.

2. The Effectiveness of Environmental Policies

In our model, we assumed that from 2014 to 2017, Shanghai would implement policies in accordance with its action plan, and that policies in the ten-year period after 2017 would continue to adhere to the general framework of this action plan, including the specific actions and measures detailed above. As such, our model simulated the effects of these actions and policies on variables such as changes in coal consumption growth, changes in emissions per unit of coal, changes in vehicle emissions/km, changes in the average annual growth in the vehicle fleet size, and impacts on the energy and transportation structures. The main results, which cover the period 2013–2027, are as follows:[6]

- Coal as a proportion of primary energy consumption will decline from the current level of 50 percent to 30 percent.
- The proportion of road travel in total transportation volume will remain around 75 percent.
- Clean coal technologies will reduce pollution emissions per unit of coal consumed by roughly 55 percent.
- Vehicle emissions/km will be reduced by 69 percent.
- Emissions intensity (emissions per unit of output) of noncoal burning sources from industry/construction will be reduced by 60 percent.
- Pollution from maritime shipping will drop by 15 percent.

3. Not Quite There

Though faithfully enforcing a slew of environmental actions and using clean coal technologies will certainly reduce emissions, our findings show that these actions alone will reduce PM2.5 by only 15 µg/m³ from the current baseline of 60 µg/m³. Even when the impact of emissions reduction in Shanghai's peripheral areas[6] is taken into account, the city's PM2.5 level will be reduced only to 42 µg/m³ by 2027, far higher than the target of 25 µg/m³.

One of the reasons for falling short is that while implementing the measures in the Shanghai action plan will reduce the city's PM2.5 level to about 45 µg/m³ by 2017, the rate at which the PM2.5 level falls in the following ten-year period (2018–2027) will slow dramatically. That is because by 2017, most of the reduction of emissions per unit of coal consumed and vehicle emissions/km will have been realized. In other words, much of the "low-hanging fruit" will have been picked (see fig. 6.2). Table 6.1 details the contributions to the overall reduction of 15 µg/m³ achieved through various existing and planned environmental actions.

III. Structural Impediments to Meeting the Target

Structural issues are the primary impediments to Shanghai's ability to meet its PM2.5 target for 2027. These factors include heavy reliance on industry, rapid growth in the size of the vehicle fleet, low level of

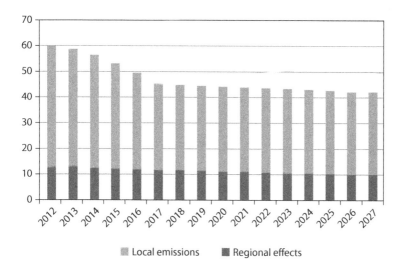

Figure 6.2 PM2.5 reductions under current and planned environmental actions (μg/m³). *Source:* Author's estimates based on PM2.5 control model.

subway travel as a proportion of urban commuting, and slow growth in clean energy as a proportion of total energy consumption. In addition, high levels of emissions from the shipping industry also pose a major challenge.

1. Overconcentration of Industry

Currently, the manufacturing industry, and particularly the heavy manufacturing industry, accounts for a relatively large portion of Shanghai's economic and industrial structure (see fig. 6.3). According to 2011 data, industry accounted for nearly 38 percent of Shanghai's GDP, and specifically, heavy industry accounted for about a quarter of its GDP.

Both of these percentages are far higher than those in other global financial centers and metropolises. For example, Singapore's manufacturing sector accounts for only around 18 percent of GDP, while less than 3 percent of London's economy is in manufacturing. Manufacturing accounts for about 10 percent of Tokyo's GDP, while the ratios are 5 percent and 3 percent for New York and Hong Kong, respectively.

TABLE 6.1
PM2.5 Reductions in Shanghai Under Current and Planned Environmental
Actions

	Overall reductions (µg/m³)	Contribution to reductions (%)
Total reductions from local efforts, including:	15	83
Emissions reduction from coal-related policies	6.5	36
- Changes in energy consumption structure	2.7	15
- Clean coal technology upgrades	3.8	21
Emissions reduction from transportation	4.3	24
- Improvements in the transportation structure	0.1	1
- Higher vehicle emissions standards	4.2	23
Emissions reduction from industrial and construction sectors	1.2	7
Emissions reductions from maritime shipping	0.9	5
Emissions reductions from policies in other areas	2.1	12
Reductions from peripheral areas	3	17

Source: Author's estimates.

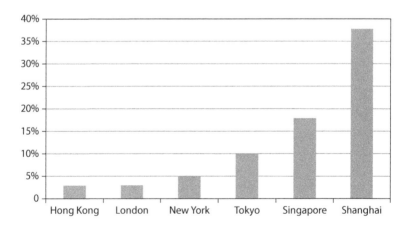

Figure 6.3 Manufacturing as percentage of GDP across global hubs, 2012. *Sources: Shanghai Statistical Yearbook* (2013) and various country and city statistical authorities; author's estimates.

It is no surprise that industrial manufacturing and the proliferation of heavy industry are responsible for much of the energy consumption and air pollution in Shanghai. In 2011, industrial coal consumption (nonpower industries) accounted for about 40 percent of total coal consumption and oil consumption in industry accounted for more than 34 percent of total oil consumption. Finally, roughly 30 percent of the city's total soot emissions and 35 percent of total SO_2 emissions came from nonpower industries.

2. *Too Many Cars on the Road*

Vehicle fleet growth must be limited to achieve the air pollution reduction goal. In Shanghai, despite the fact that a license plate auction system is already in place, the vehicle fleet size continues to grow at about 8 percent a year—far higher than the level required to hit the pollution reduction target.

Shanghai would do well to look to Singapore as an example. The Southeast Asian city-state has in place a certificate of entitlement auction system, which has been effective in controlling the growth of its vehicle fleet to within 0.5 percent a year. Currently, Singapore has approximately 1 million motor vehicles, about half of which are private cars. This translates into a private vehicle penetration rate of just 18 percent.

In contrast, the penetration rates for motor vehicles and private cars in Shanghai are both increasing rapidly and had essentially already reached Singapore's level by 2012. If annual growth in private car ownership continues at 6.5 percent in the future, the total number of private cars in Shanghai (cars licensed in Shanghai and those licensed in other localities but driven in Shanghai) will grow from the current 1.9 million to 5 million in the next fifteen years, an increase of 160 percent. This means that by 2027, Shanghai's vehicle penetration will likely be double that of Singapore.

Even assuming maximum efforts in areas such as imposing stricter standards for petroleum products, vehicle emissions, and fuel efficiency lead to a 70 percent reduction in vehicle emissions/km over the fifteen years from 2013 through 2027, the 160 percent growth in private car ownership over the same period means that PM2.5 pollution from road emissions will rise substantially rather than decline.

3. Public Transport Inadequate

Subway travel in Shanghai's central urban area accounts for just 25 percent of total commuting, compared to the average of 70–80 percent in other global financial centers. Shanghai could significantly boost its public transportation infrastructure. $PM2.5$ emissions generated by subways (in terms of transportation volume per person-km) are only one-tenth those generated by road transportation.

4. Not Enough Clean Energy

Clean energy[7] as a proportion of Shanghai's primary energy consumption is just 15 percent, much lower than that in advanced economies such as major European countries (40–60 percent), the United States (40 percent), and Japan (more than 35 percent).[8]

Although Shanghai's own municipal twelfth FYP already pledged to increase the proportion of clean energy to 21 percent by 2015, this effort, together with the adoption of currently planned environmental measures, still will not be enough to achieve the 2027 target.

5. Low Value-Added Shipping and High Pollution Intensity

Shanghai is the world leader in terms of both container throughput and port cargo throughput. From 2008 to 2012, port cargo throughput saw cumulative growth of 40 percent, according to the China Shipping Database, while growth in other international ports was just 4–5 percent. The establishment of the International Shipping Center and Free Trade Zone will further increase the total volume of international trade through Shanghai, particularly intermediary trade. As such, Shanghai's port throughput is projected to far exceed 40 million standard containers, measured in 20-foot equivalent units (TEUs), by 2020.

This volume of port traffic brings with it several pollution and energy-related issues. First, a relatively high proportion of the cargo coming through Shanghai is dry bulk cargo, which is associated with more severe pollution than container cargo. Second, container shipping is low value-added business, which means that shipping companies have to increase cargo volume to sustain thin profit margins, resulting in high growth of TEU and its associated energy consumption and pollution.

Third, water-to-water transshipment can reduce pollution, but transshipment trade as a proportion of total shipping volume is too low in Shanghai. Fourth, the use of rail transport to offset marine cargo traffic is insufficient. Cargo transport via highway generates much more pollution than rail, for example.

In terms of end-of-pipe measures, much more can be done to control the quality of petroleum products for seaborne vessels, curtail pollution from dry bulk cargo, limit tailpipe emissions of port and transport vehicles, and mitigate dust from terminal yards scattered around the ports.

IV. Necessary Structural Adjustments

It is clear that for Shanghai to meet its PM2.5 target of 25 μg/m³, major changes to the city's economic structure are required, in addition to implementing existing environmental measures. What follows are detailed discussions of the proposed structural adjustments, including policy recommendations to enable such a transition in the Shanghai economy.

1. Reduce Heavy Industry and Increase Services as a Proportion of GDP

Shanghai should aim to reduce heavy industry as a proportion of its GDP from the current 25 percent to 10 percent or less from 2013 to 2027. If such a target is met, Shanghai's energy consumption elasticity will fall by roughly 40 percent (from 0.5 to 0.3), transportation elasticity will fall from 0.8 to 0.7, and the industry and construction value-added elasticity will fall from 0.9 to 0.3, according to our estimates. Moreover, these efforts should lead to a roughly 20 percent decline in emissions intensity from heavy industry/construction. Finally, cutting the proportion of heavy industry will help reduce coal as a proportion of primary energy, which we refer to as an "indirect" adjustment to Shanghai's energy composition.

Thus, compared to the baseline scenario in which Shanghai's industry structure remains unchanged, the decline in importance of heavy industry will lead to PM2.5 reductions of 5.9 μg/m³ over the fifteen-year period.

To shift away from heavy industry, policymakers should consider various means, including substantially increasing fees for pollution emissions, restricting financing of heavy industry projects, prohibiting new heavy industry projects, and leveraging fiscal, taxation, and financial tools to incentivize the relocation of production facilities to areas that have much lower population density and far greater environmental capacity than Shanghai.

Based on Shanghai Environmental Monitoring Center data, industrial boilers and furnaces from steel producers, and industrial processes from the petrochemical industry and industrial spray painting, respectively, account for 10.2 percent and 15.4 percent of PM2.5 emissions in Shanghai. As such, consideration should be given to relocating the production facilities of Baosteel, Jinshan Petrochemical, and Gaoqiao Petrochemical, the city's largest heavy industry players, to other areas.

More specifically, an analysis of Baosteel relocation has yielded preliminary conclusions on how to disperse the company's production capacity. One, Baosteel can move a portion of its production capacity to other coastal areas, such as Shaoguan and Zhanjiang in Guangdong province. Two, the steel company can substantially increase the production capacity of its subsidiary, Xinjiang Bayi Iron and Steel in western China. Three, Baosteel can also move a portion of its production capacity to western provinces, such as Ningxia or Gansu. Four, while relocating a portion of its production facilities, Baosteel needs to simultaneously undertake industrial upgrading of its facilities to reduce pollution.

Finally, to increase the proportion of services in the Shanghai economy, the municipal government must adopt a series of fiscal, tax, and financial measures. The city should capitalize on the opportunities afforded by the newly created Shanghai Free Trade Zone (SHFTZ) to relax market access restrictions and allow more private sector and foreign investment. It would also do well to implement favorable taxation, financing, and human capital policies to promote the development of high-end services such as finance, healthcare, e-commerce, third-party logistics, IT, and consulting.

2. Reduce Vehicle Fleet Growth and Increase Subway Travel

Given the same volume, rail and subway transportation is ten times more energy efficient than automobile transportation, which means that correspondingly, pollution emissions from rail energy consumption is one-tenth that of private cars.

Based on the projection that Shanghai's total transportation volume will grow an average of 4.6 percent a year through 2027, we recommend keeping the volume of road transportation essentially at its current level, which means no growth. Moreover, we recommend limiting the annual growth in the vehicle fleet to 2 percent or less, and assume a decline of 20 percent of Shanghai's automobile utilization rate over the fifteen-year period, which is consistent with the global trend toward a long-term decline in automobile utilization. However, flat growth in road transportation volume must be accompanied by growth in rail transport volume (subway and light rail) to accommodate demand. We estimate that rail transportation volume growth will need to average 10.8 percent per year during 2013–27.

Achieving the above targets will lead to a significant increase in rail transportation as a proportion of urban resident travel in Shanghai's central areas, from the current 25 percent to 60 percent. At the same time, because road congestion will improve, thereby cutting commute time and energy consumption per unit of distance, we estimate that vehicle emissions/km will be further reduced by about 20 percent. Compared to the baseline scenario in which rail transportation as a proportion of travel remains at 25 percent, this improvement in the transportation structure will reduce the PM2.5 level by 4.5 $\mu g/m^3$ over the fifteen-year period.

In the next two sections, we discuss the policies needed to limit the size of the vehicle fleet and to incentivize the expansion of rail transportation.

3. Policy Options

I. REFORM THE LICENSE PLATE AUCTION SYSTEM The number of private vehicles with Shanghai licenses totals 1.4 million (out of a total of 2.6 million Shanghai-licensed motor vehicles). In addition, about half a million vehicles are permanently driven in Shanghai but bear out-of-province license plates.[9] Based on monthly data for private car license plate auctions, the number of vehicles with Shanghai license plates has risen by about 100,000 a year recently, while vehicles with out-of-province licenses have risen by about 50,000 a year.

It is instructive to look at Singapore in more detail to see how Shanghai might apply a better license plate auction system to curtail the growth of its vehicle fleet. As briefly discussed, Singapore has implemented a certificate of entitlement system, which was quite effective at keeping the annual growth of private car ownership to around 3 percent between 1990 and 2008. After 2008, the imposition of a stricter quota of just 5,000 license plates per year on average—a policy that has consistently received public support—led to further decline in private car growth to 0.5 percent per year. Singapore's success on this front is due to the credibility of the government and its strong enforcement capabilities, as well as continued public investment in its convenient and efficient subway network.

The advantage of the Singaporean system is that a substantial number of licenses will expire each year. Therefore, even if the annual growth of vehicle fleet size is fixed, a greater number of new licenses can be auctioned off each year, which alleviates supply scarcity.[10]

Reflecting on Singapore's experience, we believe that the Shanghai government should impose a ten-year limit for all new license plates—currently, all license plates are permanently valid—as well as requiring license plate holders to engage in competitive bidding when their plates expire. Of course, this policy may be politically challenging since holders of existing Shanghai license plates assume that their plates will be permanently valid. A sudden policy change will cause a public backlash from car owners.

Two proposals, in our view, could help ease the transition toward such a policy and create more public buy-in. One, make new licenses valid for ten years and transferrable; keep old licenses permanently valid, but prohibit them from being transferred or inherited. Or two, make new licenses valid for ten years and transferrable; change old licenses to a limited term, but also permit transfers; and extend the term of use for old licenses to slightly longer than ten years as a way to compensate current holders.

Each proposal has its merits. The first does not hurt the interests of original license holders, but will greatly reduce the number of new licenses that can be issued, which could result in a price surge when the new system is introduced. The second will avert a surge in auction prices for new licenses, but will likely create resistance and trigger complaints among holders of old licenses.

The second proposal appears to be the better option, in our view. In terms of general fairness, the costs of problems such as traffic congestion and air pollution, which have accumulated over the past decade or more, should be borne by all residents and not by a small group of new license buyers. According to the "polluter pays" principle of public finance, the majority of people should foot the bill for the pollution they have caused in the past. In addition, from the perspective of social stability, although the number of "losers" under the second proposal (more than 2 million holders of old licenses) is much larger than the potential number of losers under the first proposal (100,000 or so people a year), the per capita cost to the losers in the second proposal is far lower than in the first. Under the first proposal, only holders of new licenses will face a sudden surge in prices, which could lead to irrational behavior on the part of bidders. Under the second proposal, because the costs are spread more evenly among a much larger population, social tensions may be mitigated (see box).

How the policy could be implemented

Suppose that on the date the new policy takes effect on January 1, 2016, all newly auctioned license plates will be valid for only ten years, expiring thereafter. For license plates auctioned before that date, in order to alleviate financial pressures on license holders and accounting for the fact that auction prices have risen rapidly in recent years, a two-step method should be adopted to calculate how much longer old licenses should remain valid.

In the first step, the theoretical number of years that an old plate remains valid can be calculated with the formula: 10 + the year in which the license was won, then subtracted by 2015. For example, for a license plate won in 2010, the remaining validity is theoretically five years (10 + 2010 − 2015). In the second step, the result from step one is adjusted based on the "subsidization formula": (2 + 1.2 × the theoretical remaining number of years). That is, for that same license plate won in 2010, the adjustment puts the remaining period of validity at eight years, three more than the theoretical remaining number of years.

To summarize, changing the license plate auction system can achieve the following:

- All drivers will bear a portion of the cost due to the policy change, which will help to meet the objective of reducing congestion and pollution and adhere to the fairness principle of "all polluters pay."
- All older license plates will receive a subsidized validity period of at least two years, a grace period that offers a buffer to the increase in license plate prices.
- Considering the relatively high price of license plates in recent years, the two-step formula is designed to give more subsidized years for licenses won in recent years.
- By gradually recycling old license plates back into the auction pool, supply scarcity of license plates for auction can be managed. As a result, price surges can be avoided and relative price stability maintained.

II. INSTITUTE CONGESTION FEES   Another policy option to alleviate congestion and curb air pollution is to impose traffic congestion fees. The basic concept is to collect a certain amount from drivers in areas that experience heavy traffic and environmental stress based on price elasticity as a way to incentivize those who do not have an essential need to drive to use public transportation.

With the exception of public buses, ambulances, fire engines, police cars, and other public vehicles, all motor vehicles driven into the designated areas, particularly in downtown, should be required to pay the congestion fee. The appropriate congestion fee rates should be calculated using demand elasticity as the key method, while taking into account emissions reduction objectives.

Traffic congestion fees are essentially a price mechanism (collected from drivers) used to limit the density of vehicle flow on urban roads during peak periods. Such a fee originated in Singapore in the 1970s. At the time, Singapore also imposed electronically zoned collection of fees—three Singaporean dollars per day—on vehicles passing through a "control area" of nearly 600 hectares. The measure reduced motor vehicle traffic in Singapore by 44.5 percent, while public transportation travel increased 20 percent.[11] A similar measure was adopted in

Stockholm to reduce the number of vehicles using major arteries in the city. The electronically zoned fee in Stockholm was 60 kronor per day, or 5–20 kronor per instance for individual time segments. This measure resulted in a 20 percent reduction in traffic flow through major urban arteries.

International experience is once again instructive. London, for example, instituted in 2003 a congestion fee of £5 on vehicles entering the city center. It was a progressive fee that increased to £10 to keep pace with rising consumer purchasing power and inflation. The key to this policy's effectiveness was measuring price elasticity. The London traffic authorities used fees of £2.5, £5, and £10 in select areas to test the effect on traffic volume. They determined that a fee of £5 would achieve the objective of reducing traffic volume by 10 to 15 percent. Indeed, the policy resulted in a 12 percent reduction in traffic volume, which was consistent with the city government's projections.[12]

We estimate that Shanghai should set a congestion fee of about 20 yuan ($3.50) for smaller vehicles and a higher one for larger vehicles, calculated based on the ratio of London's congestion fee to its subway costs. Another way to determine the congestion fee is to estimate the price elasticity of traffic, and set the fee level by assuming a desired rate of traffic reduction.

Areas designated as fee collection zones should be those that have heavy pedestrian and vehicle traffic. Public transportation options should be available in these areas. In addition, an electronic monitoring system and a payment system similar to a speeding ticket system should be put in place to collect congestion fees. Drivers who fail to pay the fee should be subject to a hefty fine—in London, the fine is ten times the congestion fee.[13]

London implemented its congestion fee in 2003, resulting in revenue of approximately 68 million pounds. This figure was lower than projected, but when factors such as the reduction in traffic accidents and emissions were weighed in, the overall social benefits of the congestion fee were obvious. For Shanghai, rough calculations suggest that if a congestion fee of 25 yuan ($4) per vehicle is levied, and if on average 20 percent of the 1.4 million private cars (280,000 vehicles) pass through the designated fee collection zone, the city should be able to collect an average of 2.5 billion yuan ($403 million) a year.

III. COMBINING SEVERAL POLICIES Reforming the license plate auction system and instituting congestion fees are two options that can be implemented individually or in combination. If the license plate auction system is aggressive enough to limit the annual growth of vehicles to 2 percent or less, then it is not necessary to institute congestion fees. However, if the government decides that changing the license plate system is politically too controversial and will be met with fierce public resistance, then adopting congestion fees is the path of lesser resistance. Implementing a congestion fee alone without reforming the license plate system will effectively mean lower utilization rate (higher idle rate) of automobiles.

Of course, a less aggressive license plate auction system can be considered, such as a policy change that does not affect old license plates but only changes the term of validity for all new licenses, while a higher annual growth rate in license plates issuance of, say, 5 percent is maintained. At the same time, a congestion fee of 10 yuan ($1.60) per vehicle per day can be charged. This combination may achieve the objective of halting growth in traffic volume by reducing the car utilization rate.

4. Increase Rail Transportation

In large cities of developed countries, anywhere between 60 and 80 percent of the residents choose rail transportation for their commuting needs. To meet the target of 60 percent in Shanghai, the city must substantially increase its subway length and utilization rates.

I. EXTEND SUBWAY TRACKS Although Shanghai's total subway mileage (including the elevated maglev line from the Pudong airport) already ranks in the top two globally, its mileage per capita and mileage per unit of area are far below those of other global metropolitan areas. The Shanghai municipal government has planned for 22 subway lines totaling 800 km by 2022. Assuming that subway construction will continue to grow at the same average rate for the five years following 2022, then by 2027, total rail mileage should reach 970 km, which translates into just 0.33 km/10,000 people and 0.184 km/km². Considering the population growth potential in Shanghai, as well as the relatively rapid expansion of the city's size, it is imperative to raise the subway mileage.

We recommend raising the planned subway mileage by 230 km, to a total of 1,200 km by 2027, which translates into 0.4 km/10,000 people and 0.226 km/km², similar to the averages found in other global metropolises. The increase in mileage should primarily target routes where there is obvious need for expansion but which are not currently included in the city's plans. For example, the additional 230 km could include connector lines to subway routes 1 and 2—the city's busiest lines—to alleviate immense crowding during rush hour. It could also include building a circle line linking several downtown areas, thus reducing transfers.

II. EACH TRAIN SHOULD CARRY MORE PEOPLE Although Shanghai has already surpassed Tokyo in terms of total subway mileage, Tokyo's subway system carries some 6 million people during peak periods, compared to just 3 million in Shanghai. This is because the carrying capacity of Shanghai's subway system is below the global average for major cities.

On the Yamanote line that connects Tokyo's transportation hubs, for example, an eleven-car configuration is used. The typical configuration on New York's subway is anywhere from eight to eleven cars. Moreover, station platforms in these transportation hubs are very long, with multiple tracks, so that express and local trains can simultaneously operate. Stations along the route can also accommodate these longer trains.

Some of Shanghai's subway routes, in comparison, have much shorter platforms, requiring shorter train lengths. At present, the standard configuration on the busiest routes is only eight cars, and just six cars on less busy routes. The system's carrying capacity can be increased significantly by expanding to eleven or twelve cars on the busiest routes. This will of course require renovating stations and platforms to accommodate longer trains. Related operations and dispatching services must also be enhanced to ensure the trains' punctuality. Furthermore, carrying capacity can be increased by improving route planning and conducting simultaneous operation of express, local, and high-speed routes. The adoption of these measures will boost the Shanghai subway system's carrying capacity by about 40 percent, according to our estimates.

III. PAYING FOR NEW SUBWAY INFRASTRUCTURE To increase subway mileage by 230 km over a fifteen-year period will require an average of 9.2 billion yuan ($1.5 billion) in investment each year (not adjusted for inflation). This is based on an estimated cost of 600 million yuan (~$100 million)/km of train track. And when train length expansion and station platform renovations are incorporated, an estimated additional investment of 3 billion yuan (~$500 million) per year will be needed.

Funding for subways currently depends largely on municipal fiscal revenue and state bank loans. However, there are more novel ways to fund subway expansions. A large portion of the funding can come from the additional revenue accrued from the license plate auction fees and congestion fees, as well as from private capital and municipal bonds. Shanghai's license plate auction system can provide the municipal government with 6–7 billion yuan ($1 billion–$1.2 billion) in revenue each year. If the license plate system is reformed and new congestion fees instituted in accordance with our policy recommendations, the net fiscal revenue each year could reach 10 billion yuan ($1.6 billion), a substantial portion of which can be allocated to building new subway infrastructure.

But relying just on public spending is not sufficient and could strain budgets and lead to work stoppage. Private capital must also be harnessed in a much more significant way to accelerate subway construction and introduce competition, which should also help improve operating efficiency and service.

Hong Kong has successfully attracted private capital for public infrastructure. There, subway land-use rights and construction rights are bundled for sale. Selling zoned residential land to the subway company, along with the subway construction rights, allows the Hong Kong government to indirectly subsidize subway construction through the land price, minimizing fiscal subsidies. This approach prevents discordant construction of residential areas and subways. It also serves real estate developers' interests, because once the subway is complete, homes near the subway tend to increase in value.

In addition, rail transportation investment should be financed through municipal bonds. Shanghai should aim to lead the way in creating a municipal bond market to provide long-term financing for rail and subway projects (for example, financing with maturities from ten to thirty years). Shanghai's financial condition is among the best in the nation, and the country already has plans to develop a municipal bond market. If Shanghai is able to lead these reforms, publish its balance sheet, increase transparency of public finances, obtain the first municipal bond rating, and issue the first municipal bonds, then not only will it provide financing for infrastructure projects, it will also pave the way for developing a national municipal bond market.

With sufficient funding and highly efficient construction and operations practices, Shanghai is well positioned to boost rail transportation as a proportion of resident travel to 60 percent through 2027.

5. *Ramp Up Clean Energy*

Shanghai should aim to increase clean energy as a proportion of primary energy consumption from 15 percent to 45 percent, which includes raising natural gas to 20 percent and renewable energy to about 20 percent during 2013–27. These changes in the energy structure mean that coal as a proportion of primary energy consumption will plummet from 50 percent to 15 percent.[14] Based on our projections, compared to the baseline scenario which assumes only "indirect" adjustments to the energy composition as a result of reducing the proportion of industry, a more aggressive and direct adjustment to the energy structure will reduce the PM2.5 level by 2.6 µg/m³ over the fifteen-year period.

I. BOOST NATURAL GAS Natural gas currently accounts for about 8 percent of the primary energy consumption in Shanghai. This proportion can be increased substantially if the municipal government encourages more natural gas consumption in transport and power generation.

Indeed, the use of natural gas in seaborne shipping and public transportation is still in its infancy in Shanghai, while many northern Chinese cities have moved ahead and completely retrofitted their public transportation vehicles and heavy trucks for natural gas. According to the Shanghai Bureau of Statistics, the transportation sector accounts for about 20 percent of total primary energy consumption. In addition, energy consumed by vehicles such as buses, taxis, and trucks accounts for about 40 percent of total transportation-related energy consumption. If the Shanghai government offers modest subsidies to retrofit these vehicles to natural gas power, then gas consumption as a proportion of primary energy will rise by 8 percentage points.

In terms of power generation in Shanghai, natural gas makes up a small portion, accounting for about 13 percent of the municipality's total installed power capacity in 2010. Not only is this far below the OECD average of generating 30 percent of electricity from natural gas, it is even below the global average of 22 percent, according to the International Energy Agency. By 2027, if natural gas as a proportion of power generation in Shanghai can reach 25 percent, then natural gas as a proportion of primary energy consumption can increase by 7 percentage points.

The primary obstacle to ramping up natural gas in power generation is that it is more expensive than coal. As long as the government modestly subsidizes electricity prices, private sector investment in natural gas projects will rise. However, the current on-grid price for gas-fired power is insufficient to cover the costs of power generation. Given the importance of gas-fired power to emissions reduction and the environment, subsidies for gas-fired power should be increased.

Shanghai appears well equipped to rely more on natural gas and to implement a large-scale shift away from coal and oil. With the completion of the Shanghai trunk line of the second West-to-East Gas Pipeline and progress on phase two of the liquid natural gas (LNG) terminal project, Shanghai will have multiple sources of gas. By 2015, the second West-to-East Gas Pipeline should provide Shanghai with 2 bcm of gas each year, or about 20 percent of the natural gas supply for the entire municipality. At the same time, phase two of the LNG project will include three 165,000-m^3 storage tanks and a second submerged pipeline, thus further expanding import capacity and alleviating gas supply constraints.

Based on the analysis above, increasing natural gas as a proportion of primary energy consumption to 20 percent by 2027 appears feasible. The United Kingdom's experience underscores the feasibility of significantly increasing the use of natural gas. When the United Kingdom discovered new sources of natural gas in the 1960s, it began to widely promote the use of natural gas as a substitute for coal, oil, and other high-polluting energy sources. Consequently, natural gas as a proportion of the United Kingdom's primary energy consumption leapt from just 1 percent in 1968 to 19 percent in 1978, exceeding 40 percent as of 2008.[15]

II. DEPLOY MORE RENEWABLE ENERGY Shanghai's twelfth FYP requires renewable energy as a proportion of primary energy to reach 12 percent by 2015. At the same time, according to the Renewable

Portfolio Standard (RPS) requirements to which China has committed, renewable energy as a proportion of total energy consumption should reach 15 percent nationally by 2020. In our view, it is feasible for Shanghai to reach or exceed 20 percent renewable energy by 2027.

First, of Shanghai's current externally sourced electricity—that is, power imported from neighboring provinces—renewable energy, including hydropower and nuclear power, accounts for over 6 percent of the municipality's primary energy consumption. Shanghai had expected this figure to reach 11 percent by 2015. As China meets its national RPS targets, Shanghai's externally sourced renewable energy should constitute 15 percent of the city's primary energy consumption.

Second, Shanghai has national support to prioritize wind power development. As a coastal city, Shanghai has rich offshore wind resources, estimated at more than 33 GW. At the end of 2010, Shanghai's installed capacity of wind power totaled 210 megawatts (MW), with installed capacity expected to reach 1 GW by 2015.

Despite wind power's higher investment and generation costs, sustained government subsidies should bring installed capacity up to 10 GW by 2027. By then, wind power generation capacity will have reached 28 billion kWh (based on 2,800 hours/year), which will raise wind as a proportion of primary energy consumption to 4 percent.

Third, solar power generation holds great potential as it is currently starting from a very low base—for instance, installed capacity at the end of 2010 was only 20,000 kW. Judging from Germany's experience, government subsidies are crucial to incentivizing solar power generation. The German government used long-term, low-interest loans to encourage R&D and construction of photovoltaic power generation. It also set higher on-grid tariffs to incentivize private investment in solar power projects. As a result, solar accounted for 7.5 percent of total power generation capacity in Germany in 2015.

China's current subsidy for grid-connected solar power is only about half that of Germany's. Therefore, we recommend that Shanghai bolster subsidies for R&D and deployment of solar power generation so that by 2027, total installed solar capacity can reach 2 GW, averaging 33 percent annual growth. If that is achieved, we project solar as a proportion of Shanghai's total primary energy consumption to approach 1 percent by then.

III. WHAT SPECIFIC MEASURES ARE NEEDED? Several policies can support Shanghai's adoption of renewable energy:

- Tax conventional coal to increase its cost. This can be done by hiking the resource tax on coal. At the same time, Shanghai can lead the way by raising fees for SO_2 and NOx emissions.
- Substantially increase subsidies for clean energy. National subsidies for renewable energy at present are about 0.2 percent of GDP, and additional subsidies from the Shanghai government account for about 0.05 percent of GDP. In the United States and Germany, these figures are 0.4 percent and 0.7 percent, respectively. To boost renewable energy in the future, the government must increase subsidies, for example, by using a portion of pollution fees to subsidize clean energy. Shanghai should increase local subsidies to invest in and maintain clean energy vehicles and equipment, and to further subsidize electricity generation from renewable energy.
- Explore a clean energy voucher trading system. Power companies that exceed targets for generating electricity from renewable energy sources can sell their vouchers to companies that fail to meet targets. This can form a cross-subsidization mechanism in which companies that fail to meet targets help to subsidize those that succeed in meeting standards.

6. Curtail Emissions in the Shipping Industry

Shipping-related atmospheric emissions in Shanghai, such as those from ports, vessels, and stockyards, should be cut by 50 percent over the fifteen-year period. Compared to the baseline scenario of just a 15 percent reduction in maritime shipping emissions, achieving this more aggressive reduction target can reduce the PM2.5 level by 2.0 µg/m³ in Shanghai, according to our projections.

I. SHIFT TO HIGH VALUE-ADDED SERVICES AND BOOST CARGO TRANSSHIPMENT Compared to established shipping hubs such as London, Hong Kong, and Singapore, the volume of transshipment cargo, the proportion of container cargo, and the level of development of high-end shipping services in Shanghai are much lower. These structural issues all contribute to pollution emissions. With the momentum behind the creation of the SHFTZ, the port of Shanghai should focus on developing high-end services and significantly increase the proportion of transshipment cargo.

Our specific recommendations are as follows:

- Shanghai should deliberately exit the low-end shipping business, particularly pollution-intensive bulk cargo shipments. It should no longer view raising port throughput (tons) as a primary target and should avoid engaging in cutthroat competition with neighboring ports for low-end logistics business. The focus should not be on volume but on shifting toward higher value-added logistics services.
- Drawing on the experiences of London, Hong Kong, and Singapore, Shanghai should seek to elevate its status as a transshipment port and strive to become a transshipment center for northeast Asia. At a minimum, this will require reforms to simplify delivery and transaction procedures, and will require the adoption of an aggressive and flexible pricing strategy, the provision of shipping financing and marine insurance, and the establishment of internationally recognized maritime arbitration. All of these actions should create an environment that attracts talent specializing in high-end logistics, legal services, financial services, and insurance, which would in turn attract multinational corporations to set up international distribution centers in the SHFTZ.
- Using Singapore's experience as an example, Shanghai should take full advantage of its strength in shipbuilding and maintenance, shipping equipment, maritime support, and services, and should prioritize applying new technologies to maritime shipping. It should attract investors and risk capital into the areas of maritime shipping as well as coastal and oceanographic engineering. In the policy arena, Shanghai should aggressively attract high-end manufacturers to set up operations in the SHFTZ and promote higher-value trade.
- As part of the planning for logistics infrastructure, Shanghai should focus on extending rail transportation to major deepwater ports in order to substantially increase the proportion of cargo

distributed by rail routes. Based on international experience, Shanghai has the potential to increase the volume of rail cargo by a factor of ten or more. Substituting rail for highway routes in the distribution of ship cargo will significantly reduce pollution generated in the transportation process.

II. ENCOURAGE FUEL SWITCHING AND ENGINE UPGRADES Air pollution from shipping vessels accounts for over 90 percent of total shipping-related PM2.5 emissions.[16] While seaborne vessels have higher tonnage and a greater displacement volume, meaning that they generate the greatest volume of PM2.5 emissions, the primary cause of ship-related PM2.5 emissions is the use of low-quality petroleum products. (Atmospheric pollutants from river vessels are relatively minor overall, but they have a larger impact on urban areas because these vessels travel in closer proximity to cities.) Most vessels use fuel at the lower end of limits specified by the International Maritime Organization (IMO).

Therefore, we recommend that Shanghai aggressively implement fuel switching for ships when they reach port, establish emissions control zones, encourage ships to rely on onshore power while docked in port, and replace oil with natural gas/LNG for powering mechanical equipment in the piers.

The "Fair Winds Charter"[17] that Hong Kong implemented in January 2011 offers important lessons for Shanghai in encouraging the switch to low-sulfur fuel. After Hong Kong implemented the fuel change agreement, shipping-related SO_2 dropped 80 percent and PM2.5 emissions also fell significantly.[18] Furthermore, as a result of implementing the regulations on shipping emissions, total SO_2 and PM2.5 emissions in Hong Kong dropped 21 percent from the 2008 level.[19] Companies that voluntarily signed the Hong Kong fuel agreement and companies that did not sign were required to bear different operating costs, with the signatories receiving a waiver of 60 percent of port facility and lighting fees. However, companies still had to assume anywhere between 50 and 80 percent of the fuel change cost, which was determined based on the shipping rating, vessel type, and displacement volume of the vessel.

Shanghai should thus reflect on Hong Kong's experience and implement a fuel change agreement that is initially valid for two years.

The city should use the comparative advantages of SHFTZ to present an agreement similar to the one contained in the Fair Winds Charter, which requires ships entering port to change their fuel to one in which the sulfur content is no more than 0.5 percent, and switching back to regular fuel when they depart. Like Hong Kong, Shanghai should also make signing the agreement voluntary.

The IMO will also be establishing emissions control areas (ECAs) in certain major port cities to coordinate controlling vessel emissions across different regions. While maintaining the 0.5 percent low-sulfur fuel change policy, Hong Kong's 2013 Fair Winds Charter (the charter is reaffirmed annually) also called for enhancing cooperation in the PRD region to ensure consistency in enforcing mandatory emissions regulations issued at the end of 2013. Based on the implementation of the fuel change plan, Shanghai should also establish a centralized ECA for the YRD region to deal with the sulfur content of port vessels.

Currently, Asia does not have a designated ECA.[20] Shanghai can lead the coordination effort in the YRD and establish Asia's first ECA there. According to the provisions of the IMO Annex, the sulfur content of fuel used in ECAs should not exceed 0.1 percent. In the Shanghai port area, SO_4^{2-} (sulfate) partly formed by SO_2 is an important component of PM2.5 emissions.[21] Therefore, by lowering the sulfur content from the current 3.5 percent to 0.1 percent, emissions from shipping vessels can be reduced by more than 10 percent.

Drawing on the experiences of Europe and North America, the establishment of an ECA in the YRD can also be used to upgrade and retrofit the engines of 90 percent of vessels entering and leaving the port of Shanghai by 2027. While Shanghai should aggressively promote engine upgrades and retrofits of Chinese river and seaborne vessels to achieve dual fuel capability, only 4.2 percent of the vessels that dock in Shanghai are Chinese. Therefore, Shanghai should encourage international seaborne vessels to use hybrid engines and cleaner fuels, in particular to boost the proportion of LNG to 40 percent, based on the Anchorage fuel change policy that is similar to the Hong Kong charter.

In fact, the Ministry of Transportation has already begun to upgrade vessels to use cleaner fuels, and the pace at which ships are turning to LNG is accelerating. Hybrid vessels can substitute LNG for 50 to 80 percent of diesel fuel and still maintain stable operations. In addition,

China has already developed its own LNG/diesel hybrid vessel engine technology.

Because vessel engine technology continues to advance, it is possible that 30 percent of river vessel engines may be retrofitted by 2020 and 70 percent by 2027, which can help reduce shipping-related emissions by nearly 10 percent. As a result, by 2027, engine upgrades of international seaborne vessels and additional adoption of hybrid engines can help reduce shipping-related emissions by more than 15 percent, according to our estimate.

III. SWITCHING FROM GASOLINE TO LNG AND GASOLINE TO ELECTRICITY Trucks transporting shipping containers in the port area are a significant source of emissions. Implementing an LNG container truck policy can reduce port emissions. For instance, the Ningbo port is already piloting a program to replace traditional vehicles with LNG container trucks. This not only reduces emissions, but also has tangible economic benefits. By using LNG container trucks, the savings per vehicle can total as much as 100,000 yuan ($16,000) per year.[22]

We recommend a pilot program to introduce 500 LNG container trucks in the Shanghai port area in 2015, with the government offering subsidies of 20,000 yuan ($3,500) per container truck and 30,000 yuan ($5,000) per tanker truck. By 2017, half of the container trucks should use LNG, with full replacement by 2025.

Mechanical operating equipment is also a source of emissions in the port area, accounting for over 3.5 percent of total port emissions.[23] Changing the power source of port equipment to LNG-generated electricity can curtail PM2.5 emissions. Considering that Shenzhen has already confirmed that it will adopt a fuel change policy and complete all equipment conversions from gasoline to natural gas and electricity in its port area, Shanghai should follow suit and accelerate the pace of LNG container truck replacement and fuel conversion of operating equipment. We estimate that by 2027, the conversion of mechanical equipment from gasoline to natural gas and electricity will help reduce shipping-related PM2.5 emissions by about 3 percent.

IV. PROMOTE THE USE OF ONSHORE POWER The emissions generated while docked vessels are loading or unloading cargo account for approximately 10 percent of total shipping-related PM2.5 emissions. Currently, vessels in Shanghai rely primarily on auxiliary generators and steam boilers to provide basic power supply to maintain operations while moored. Encouraging the use of onshore power will significantly reduce vessel emissions as they sit in port.

After completing onshore power infrastructure at Yangshan Harbor and the Wusong International Cruise Terminal in 2015, Shanghai should complete the construction and renovation of onshore power facilities at all port areas under its jurisdiction. While building onshore power facilities, the unique features of various vessel types must be considered to accommodate the electrical loads of different vessels and improve cable management systems. LNG-generated power must be used to minimize pollution to the greatest extent possible. Given the rapid growth in port throughput, the full application of onshore power technology is estimated to reduce shipping-related emissions by more than 5 percent.

If the growth in shipping and port throughput does not exceed 30 percent, as long as Shanghai effectively adjusts its shipping industry and adopts a series of measures to control emissions, Shanghai should be entirely capable of cutting shipping-related emissions in half through 2027.

V. STRUCTURAL ADJUSTMENT'S IMPACT ON PM2.5 Of the 35 µg/m³ reduction required for Shanghai to realize its PM2.5 target—reduction from 60 µg/m³ to 25 µg/m³ in fifteen years—planned environmental measures and emissions reductions in peripheral areas will contribute 20 µg/m³, while structural adjustment policies will contribute the additional 15 µg/m³ needed to achieve the overall target (see fig. 6.4). The specific emissions reduction results achieved via different structural adjustment policies are shown in table 6.2.

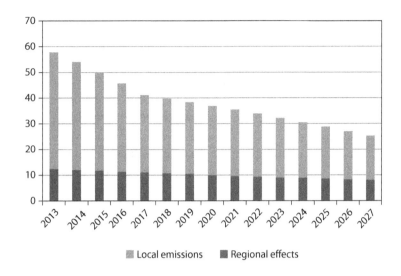

Local emissions ■ Regional effects

Figure 6.4 PM2.5 reductions under structural adjustment actions (μg/m³). *Source:* Author's estimates based on PM2.5 control model.

TABLE 6.2
PM2.5 Reductions in Shanghai under Structural Adjustment

	Overall reductions (μg/m³)	Contribution to reductions (%)
Reductions from environmental measures	15	43
Reductions from structural adjustment, including:	15	43
1) Industry	5.9	16.9
2) Highway and rail transport	4.5	12.9
3) Energy composition	2.6	7.4
4) Shipping	2.0	5.7
Reductions in peripheral areas	5	14
Total reduction in fifteen years	**35**	**100**

Source: Author's estimates based on PM2.5 control model.

CHAPTER SEVEN

Case Study: Beijing

Introduction

As the political, economic, and cultural center of China, Beijing must meet the national PM2.5 emissions reduction target as quickly as possible. Alarmingly high levels of PM2.5 have gravely affected the image of the Chinese capital and the health of its people. The need to address pollution is urgent: we believe that, from 2014 to 2028, Beijing should aim to reduce its annual PM2.5 emissions from an average of 90 μg/m³ to 35 μg/m³, which is the WHO's recommended interim target-1 standard.[1] While this target may seem aggressive for Beijing, readers should note that the WHO's AQG set the PM2.5 guideline at just 10 μg/m³. In other words, the 35 μg/m³ goal is still 250 percent higher than the level the WHO deems safe. In our view, Beijing has the capacity to achieve this objective (35 ug/m³ by 2018) for the following reasons.

By various measures, such as per capita income, fiscal and financial resources, and policy environment, Beijing's capacity to reduce pollution significantly exceeds the national average. The Chinese capital had already reached a per capita GDP of 99,995 yuan ($16,278) in 2014, according to the Beijing Municipal Statistics Bureau and the National Bureau of Statistics.[2] Not only is that far above the national average per capita GDP, it puts Beijing at a level comparable to upper-middle income countries designated by the World Bank (see figure 7.1). Therefore, relative to other municipalities, Beijing has a better economic foundation to achieve the necessary emissions reduction.

In addition, concerted efforts as part of integrating the Beijing-Tianjin-Hebei region (the "Jing-Jin-Ji") will increase the emissions reduction potential for Beijing. As figure 7.2 shows, Jing-Jin-Ji is more

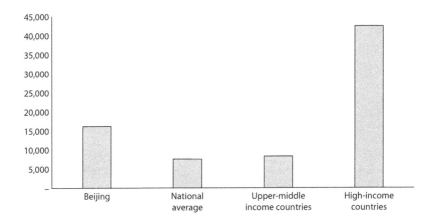

Figure 7.1 Per capita GDP comparison, 2014 (USD). *Sources: China Statistical Yearbook;* author's estimates.

Note: This includes upper-middle income countries and high-income countries as defined by the World Bank.

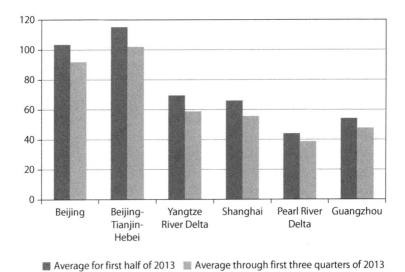

■ Average for first half of 2013   ■ Average through first three quarters of 2013

Figure 7.2 Average PM2.5 concentrations in key regions and cities, 2013 (μg/m³). *Source:* MEP.

severely afflicted by pollution than the YRD and PRD regions. In fact, as much as 36 µg/m³, or 40 percent of Beijing's average PM2.5 emissions in 2013, resulted from spillover effects from peripheral areas. Regional coordination of pollution control will boost Beijing's emissions reduction potential. In particular, if an interregional compensation mechanism is adopted, the total cost of emissions reduction can be significantly lowered or, given the same amount of spending, would improve the potential for emissions reduction.

An examination of the experiences of major global cities such as London, Los Angeles, and Tokyo suggests that it is feasible for Beijing to hit its 35 µg/m³ emissions target by 2028. The data from these cities demonstrate that notable improvement in air quality can be achieved over fifteen to twenty years (see figure 7.3).

I. Adapting the PM2.5 Control Model for Beijing

As we did for Shanghai, we built a PM2.5 control model specifically for Beijing to quantify the effects of environmental measures and structural adjustments on reducing the PM2.5 level.[3] Based on the results, we concluded that environmental actions alone would not be sufficient

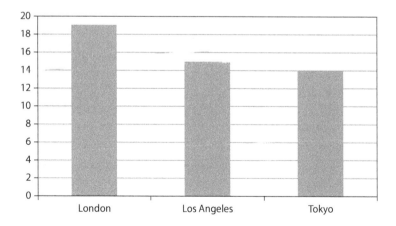

Figure 7.3 Years required for air quality improvement in major global cities. *Source:* Author's estimates.

Note: Air quality improvement means reducing major pollutants by 60 percent or more.

to achieve the PM2.5 emissions reduction target of 35 μg/m³ by 2028. We used the model to simulate a series of structural adjustment measures needed to achieve the target.

The key assumptions used in the model for Beijing are as follows:

- **Target:** Reduce average concentrations of PM2.5 from 90 μg/m³ to 35 μg/m³ by 2028.
- **GDP growth:** Real GDP growth will gradually slow from 7.7 percent in 2013 to 4.6 percent in 2028, averaging 5.7 percent during the fifteen-year period.
- **Coefficients:** Based on historical data, Beijing's average energy elasticity coefficient is 0.48 and its transportation elasticity coefficient is 0.6. (In the simulation of the "planned environmental measures only" scenario, we assumed that these coefficients would remain unchanged. In the "structural adjustment" scenario, we adopted lower elasticity coefficients for both.)

We also derived additional estimates used in our model simulations:

1. Based on the assumptions above, and not accounting for changes in the industry and transportation structures, energy consumption and transportation volume growth from 2014 to 2028 are estimated to average 2.9 percent and 3.9 percent, respectively.
2. Using data from China's MEP and Greenpeace, we estimated that coal burning contributed to about 17 percent of PM2.5, transportation 22 percent, noncoal emissions from construction and industry (including VOCs and dust) 17 percent, interregional spillover 40 percent, and other sources the remaining 4 percent. These percentage contributions served as the basis of our simulations (see figure 7.4).
3. Desulfurization, denitration, and dust removal efforts are estimated to cut pollution emissions per unit of coal consumed by 35 percent. Improvements in petroleum product quality, fuel efficiency, and automotive emissions standards can reduce average vehicle emissions by 64 percent.

Given the assumptions and estimates, even if emissions reductions from the various environmental measures are fully achieved, the 2028 target of

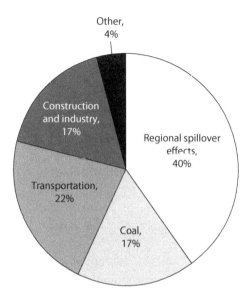

Figure 7.4 Major sources contributing to PM2.5 in Beijing. *Sources:* MEP's popular science handbook; Greenpeace analysis; relevant literature.

35 µg/m³ still cannot be met. Additional efforts are required, primarily in terms of altering Beijing's industry, energy, and transportation structures. In addition, we recommend the establishment of a regional emissions control mechanism to coordinate with Beijing's peripheral areas.

The package of policies and actions we propose considered various factors, including social acceptance, international experience, the availability of natural resources, and technical feasibility.

· II. Environmental Measures: Necessary but Insufficient

1. Brief Overview of Beijing's Existing Environmental Policies

On August 23, 2013, the municipal government released its "Beijing Clean Air Action Plan (2013–2017),"[4] stipulating that "by 2017, annual average concentration of PM2.5 will be reduced by 25 percent or more in the municipality, to about 60 µg/m³." As part of this plan, Beijing's current and planned environmental actions to curb atmospheric pollution can be categorized in the following areas:

- Control coal-related pollution and industrial coal burning;
- Optimize the energy structure and deploy clean energy substitutes;
- Curb industrial emissions;
- Control motor vehicle emissions;
- Expand the public transportation system;
- Establish comprehensive dust controls; and
- Enhance household waste disposal methods.

2. The Effects of These Actions

To evaluate the overall effects of the environmental actions through 2028, we assumed in our control model simulations that policies after 2017 would continue to adhere to the general concepts contained in the Clean Air Action Plan. These include continued negative growth in coal consumption, stricter emissions standards and petroleum product standards for new vehicles, and staying the course on VOC and dust control efforts. If the enforcement intensity of these policies is "ordinary"—that is, if 85 percent of the emissions reduction target is achieved—and this level of enforcement is maintained into the future, the effects of these proposed environmental policies through 2028 are as follows (see below for explanation of the enforcement intensities).

I. COAL AS A PROPORTION OF ENERGY CONSUMPTION WILL DECLINE FROM 25 PERCENT TO 8 PERCENT Beijing currently consumes about 23 million tons of coal, accounting for about a quarter of its total energy consumption. According to the municipal government's policy documents, such as "Breakdown of Major Tasks in the 2013–2017 Clean Air Action Plan"[5] and the "2013–2017 Work Plan for Accelerated Reduction of Coal Burning and Development of Clean Energy,"[6] Beijing is required to cut coal consumption by 8 million tons by 2015 from 2012 levels and to cut an additional 1.3 million tons from 2015 to 2017. This should reduce coal as a proportion of primary energy consumption to 10 percent or less by 2017.

Because the proportion of coal will already be quite low by 2017, the scope of further reduction is limited. Therefore, we estimate that the proportion of coal in primary energy consumption can be reduced to about 8 percent, while clean energy will rise to just over 50 percent by 2028.

II. ROAD TRANSPORT WILL FALL BY 10 PERCENTAGE POINTS Road transportation currently accounts for about 83 percent of total traffic in Beijing's central urban areas. Based on the "Beijing Urban Rail Transportation Construction Plan (2011–2020),"[7] Beijing will have 30 rail lines totaling 1,050 km by 2020. If all current and planned projects are built, we estimate that road transportation volume in Beijing's central urban areas will decline to 73 percent and rail transportation will rise to 27 percent by 2028. However, Beijing's proportion of rail transportation will still be notably lower than that of other major global cities.

III. SINGLE VEHICLE EMISSIONS WILL BE CUT BY NEARLY TWO-THIRDS On February 1, 2013, Beijing began enforcing the "Beijing V" auto emissions standard (equivalent to the National V standard) on new light vehicles, no longer permitting sales or registration of vehicles that do not meet this stringent standard. Single vehicle emissions will be reduced by about 40 percent compared to the National IV standard. It will take about ten years for all vehicles to comply with the Beijing V standard.

Conforming to the Beijing V emissions standard should lead to a 40 percent drop in NOx emissions for light vehicles and heavy diesel-powered vehicles, according to estimates from the Beijing Environmental Protection Bureau.[8] Moreover, NOx emissions will fall by an average of 15 percent when the Beijing V gasoline standard is phased in.

Stricter automobile emissions standards, higher petroleum product quality, and improved fuel efficiency should lead to an average of 64 percent reduction of emissions/km for all vehicles through 2028, according to our estimate.

IV. EMISSIONS PER UNIT OF COAL CONSUMED WILL DROP BY ABOUT ONE-THIRD By 2017, Beijing plans to reduce the sulfur content of coal from 0.5 percent to 0.4 percent and reduce the ash content from 13 percent to 12.5 percent, in large part by expanding the deployment of desulfurization technology. In addition, the Beijing Municipal Commission of Rural Affairs in 2013 led the way in formulating a "Coal Reduction and Replacement and Clean Air" action plan that aimed to slash the consumption of low-quality coal by 8 million tons per year. All districts and counties were required to cut back burning inferior coal by 20 percent or more from 2012 levels. We estimate that, due to the application of clean coal technologies and the cutback in the consumption of inferior coal, the cumulative decline in emissions per ton of coal over the years 2014–2028 will be about 35 percent.

V. NONCOAL EMISSIONS FROM INDUSTRY/CONSTRUCTION WILL BE CUT BY MORE THAN HALF Noncoal emissions primarily include construction and industrial dust and VOCs. Beijing has already mandated slashing the volume of dust by 20 percent and 30 percent by 2015 and 2017, respectively. It also required cutting VOC emissions by 10 percent each year from 2013 to 2017. By 2017, the cumulative reduction in noncoal emissions will be about 50 percent from the 2012 levels.

Dust emissions primarily come from over 150 million m² of construction sites, and VOC emissions from automotive, furniture spray painting, and certain manufacturing operations. Their respective contributions to PM2.5 are roughly equal. Because the potential for emissions reduction will decrease, the annual rate of reduction will likely fall sharply after 2017. Thus, we estimate that the cumulative reduction per unit of output from industry/construction will be about 55 percent through 2028.

3. These Policies Alone Will Not Get Beijing to the Finish Line

Based on the results of our model simulation, we concluded that if all actions are successfully executed under the "planned environmental measures" scenario, and if such actions are sustained, Beijing's PM2.5 level will decline to just 60 μg/m³. Even when PM2.5 reductions in peripheral areas are considered, the emissions level in Beijing can only be reduced to 45 μg/m³ by 2028, still significantly higher than the target of 35 μg/m³.

An important caveat is that enforcement in Beijing will likely be more difficult than in other regions for several reasons. One, Beijing's reduction target is more aggressive than that of other regions, partly due to political pressure on the capital city. In reality, these measures

TABLE 7.1
Beijing's PM2.5 Reductions Under Various Enforcement Intensity Levels

Enforcement intensity	2028 PM2.5 level ($\mu g/m^3$)
Strongest	45.1
Ordinary	52.1
Weaker	57.1

Source: Author's estimates.

may face stronger resistance in Beijing than in other regions. Two, with more than 5 million cars on Beijing roads, air pollution from road transportation is more difficult and costly to address than industrial, construction, or shipping-related emissions.

Therefore, we considered three levels of enforcement intensity: strongest = all targets achieved; ordinary = 85 percent of targets achieved; and weaker = 75 percent of targets achieved (see table 7.1). The 45 $\mu g/m^3$ result above was based on the assumption that the Beijing air pollution plan through 2017 is strictly enforced and that the peripheral areas also achieve significant reductions ("strongest" enforcement level). Table 7.2 breaks down the effects of various

TABLE 7.2
Beijing's PM2.5 Reductions Under "Ordinary Enforcement"

	2013 ($\mu g/m^3$)	2028 ($\mu g/m^3$)	Amount of reductions ($\mu g/m^3$)	Contribution to reductions (%)
Peripheral areas	36.0	21.9	14.1	37
Local emissions	54.0	30.2	23.8	63
Coal-related	15.3	3.6	11.7	31
Transportation-related	19.8	11.7	8.1	22
1) Road	16.8	8.6	8.2	22
2) Other transportation	3.0	3.0	−0.1	0
Construction and industrial	15.3	11.9	3.4	9
Other	3.6	3.1	0.5	1
Total PM2.5 level	**90.0**	**52.1**	**37.9**	**100**

Source: Author's estimates based on PM2.5 control model.

environmental actions on emissions reduction under "ordinary" enforcement intensity, which is a more likely outcome in our view.

III. Challenges in Meeting Beijing's Target

Regional and structural issues handicap Beijing's ability to meet its PM2.5 target. That is, Beijing gets much of its pollution from surrounding areas that are poorer and have fewer resources to tackle their local pollution. A 2013 Greenpeace study of pollution sources around Beijing found that southerly and westerly winds, combined with other factors, caused Beijing's PM2.5 level to rise by 16 μg/m³.⁹ In other words, slow-moving, low-pressure systems from Hebei and Shandong provinces that blow significant amounts of pollutants into Beijing are major contributors to the dense smog that continuously envelops the capital city.

In terms of structural factors, Beijing, like Shanghai, is beset by rapid growth in its motor vehicle fleet, low proportion of resident travel by subway, and a high percentage of industry in its local economy.

1. Peripheral Area Woes

I. POLLUTION AND EMISSIONS REDUCTION PLANS AROUND BEIJING Five provinces and municipalities sit on Beijing's periphery: Tianjin, Hebei, Shanxi, Shandong, and the Inner Mongolia Autonomous Region. Tianjin, along with six cities in Hebei province, were among the ten Chinese cities with the worst air quality in 2013 (see figure 7.5).

On September 17, 2013, six agencies, including MEP and NDRC, jointly issued "Detailed Regulations for Implementing the Atmospheric Pollution Prevention and Control Action Plan for the Beijing-Tianjin-Hebei Region and Peripheral Areas,"¹⁰ which laid out the 2017 atmospheric pollution control targets for the Jing-Jin-Ji region and overlaps with Beijing's 2017 air pollution action plan.

Some of the specific targets included reducing the average annual PM2.5 level in Beijing to 60 μg/m³; cutting PM2.5 levels by 33 percent from 2012 levels in six peripheral cities (Shijiazhuang, Tangshan, Baoding, Langfang, Dingzhou, and Xinji); and reducing PM2.5 levels

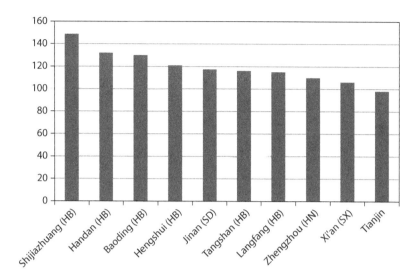

Figure 7.5 Ten Chinese cities with highest PM2.5 levels in 2013 (μg/m³). *Source:* www.PM2d5.com.

Note: HB = Hebei Province; SD = Shandong Province; SX = Shaanxi Province; HN = Henan Province.

by 30 percent or more in Xingtai and Handan, 25 percent or more in Qinhuangdao, Cangzhou, and Hengshui, and 20 percent or more in Chengde, Zhangjiakou, and other cities with relatively severe pollution. While targets were numerous, no effective regional coordination mechanism for curbing pollution was proposed.

II. TACKLING POLLUTION AND DISRUPTING THE HEBEI ECONOMY Beijing's neighboring province Hebei has some of the worst PM2.5 pollution in the country. Based on an MEP report on air quality in seventy-four cities, seven of the ten cities with the highest average PM2.5 levels in 2013 were located in Hebei: Xingtai, Shijiazhuang, Baoding, Handan, Hengshui, Tangshan, and Langfang. Among these, Xingtai had the highest average annual PM2.5 level, at 155.2 μg/m³. Cognizant of the province's environmental problems and its tarnished image, the Hebei government in 2013 adopted an emissions reduction plan and announced its implementation.[11]

The plan proposed to cut steel production capacity by 60 million tons from 2013 to 2017. It also required environmental impact assessments to be conducted for all new construction projects and expansion and renovation projects, and it placed strict approval controls on projects in highly polluting and energy-intensive industries in environmentally vulnerable areas. Moreover, the plan initiated a program to relocate or renovate firms in high-polluting industries including steel, petrochemical, nonferrous metallurgy, cement, and plate glass, so that they would be far removed from urban centers. By 2017, 123 pollution-intensive firms would be relocated, and approvals would be withheld for projects that seek to expand capacity in industries such as steel, cement, electrolytic aluminum, plate glass, coking, nonferrous metal, calcium carbide, and iron alloy.

In addition, special limits on emissions were imposed on new construction projects and coal-burning boilers of high-polluting producers in six cities: Shijiazhuang, Tangshan, Langfang, Baoding, Dingzhou, and Xinji. Meanwhile, special limits on atmospheric pollutants were applied to the thermal power, steel, and cement industries in Xingtai and Handan. Hebei also took a number of practical steps, including closing 8,347 small, high-polluting companies in industries such as oil refining, tanning, and electroplating industries, as well as suspending or adjusting operations in heavy industries with excess capacity.[12] All of these measures should improve air quality in both Hebei and the Jing-Jin-Ji region.

However, if the Hebei government alone assumes the costs of these pollution mitigation efforts, it will place an enormous burden on the local economy and its fiscal revenue. Secondary industry accounted for 52.7 percent of Hebei's GDP in 2012, and high-polluting firms such as steel and chemical plants remain the pillars of the local economy. Therefore, curbing heavy industry could significantly hurt local economic growth for several reasons.

First, achieving the pollution targets to which the Hebei government has committed itself could well lead to substantial economic and fiscal revenue declines and rising unemployment. According to a Dragonomics report,[13] the Hebei government's intention to cut steel capacity by 60 million tons would cost it about 90 billion yuan ($15 billion) a year, result in a 10 percent reduction in fiscal revenue, or 16 billion yuan ($2.75 billion), and see the loss of 200,000 jobs.

Second, enormous subsidies are needed to support Hebei's transition to clean energy sources, and the provincial government will be hard-pressed to support these structural adjustments while the economy is shrinking and public finances deteriorating. These factors underscore the reality that if the Hebei government goes it alone on emissions reduction, it will face an economic downturn and unemployment in the process. Therefore, addressing PM2.5 emissions will quickly become a political problem as the government meets substantial public resistance.

But the Hebei government does not deserve all the blame for the severe pollution in the province. Part of the problem is the legacy of the central government's decision under Mao Zedong regarding the geographical allocation of industries based on the national development strategy of "heavy industry first." The decision was made to place a large number of heavy industries in Hebei because of the province's wealth of coal resources, the existence of a power industry infrastructure, and abundant cotton cultivation. In a sense, Hebei assumed enormous costs associated with heavy industry and the resulting pollution as it supported national development goals historically.

Because about 40 percent of Beijing's current PM2.5 comes from peripheral areas, the capital will benefit from emissions reduction in Hebei. If Beijing invests in Hebei's emissions reduction efforts, it will help curb pollution in Hebei while ameliorating pollution in Beijing, and it will accomplish this more efficiently than it would by investing in mitigation efforts in Beijing alone.

In sum, it is neither realistic (from the perspective of Hebei's own financial condition) nor fair (from the perspective of the "beneficiary pays" principle) for Hebei to assume all costs of local emissions reduction efforts. Specific recommendations on how to address this challenge will be discussed later in the chapter.

2. Too Many Cars, and Growing Too Fast

The "Breakdown of Key Tasks in the Beijing Clean Air Action Plan (2013–2017)" stipulated that Beijing's motor vehicle fleet had to be limited to 6 million vehicles by 2017. Meanwhile, Beijing introduced a series of policies to limit car purchases and established stringent rules on registering new vehicles for civil servants in the central government. Despite the promulgation of these policies, the growth of the city motor vehicle fleet remains stubbornly high.

Under the policies currently in force, Beijing's vehicle fleet size will continue to grow at an average of 4 percent a year through 2020, after which growth will dip slightly to 3 percent, according to our estimates. At these rates, Beijing's vehicle fleet is expected to grow from 5.2 million in 2013 to 7.4 million in 2028, a trend that will not benefit air pollution mitigation.

3. Not Enough Public Rail Transportation

Based on existing data, commuting by subway in Beijing's central areas accounts for only about 17 percent of total resident travel.[14] This figure is around 25 percent in Shanghai, and closer to 70–80 percent in many major international cities. Given Beijing's public transportation plans and the constraints on local finance, we estimated the proportion of subway travel in Beijing's central areas to increase by 10 percentage points to 27 percent in 2028. Still, this projection appears to be relatively conservative compared to those of Shanghai and other global cities.

Beijing's low proportion of subway travel has to do with the relatively low subway mileage per capita. At the end of 2012, the per capita mileage figures were 0.49 km/10,000 people in London; 0.45 km/10,000 people in New York; 0.43 km/10,000 people in Taipei; and only 0.21 km/10,000 people in Beijing (see figures 7.6 and 7.7).

According to the Beijing municipal government, plans for building a subway network of "three circle lines and four horizontal, five vertical, and seven radiating lines" will have been completed in 2015, at which point total operating length will exceed 660 km.[15] Furthermore, the "Beijing Urban Rail Transportation Construction Plan (2011–2020)" stated that by 2020, Beijing's rail transportation network will include 30 lines and a total length of approximately 1,050 km. Assuming a permanent resident population of 24 million in 2020, we estimated that per capita subway mileage in Beijing will reach 0.4 km/10,000 people, approaching the level found in major global cities.

Even if this subway expansion target is achieved—and assuming that maximum effort had been applied in all other areas of emissions reduction—total transportation-related PM2.5 emissions will decline from the 2013 level of 19.8 μg/m³ to only 11.7 μg/m³ in 2028. As a result, the overall PM2.5 level for Beijing will still be significantly higher than the target of 35 μg/m³.

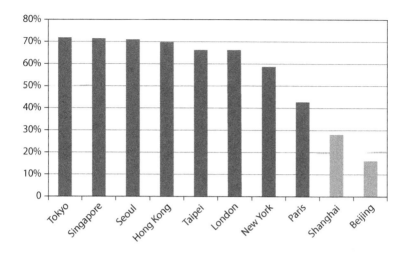

Figure 7.6 Rail transport as a proportion of total travel in global cities. *Sources:* Author's estimates; public sources from various countries.

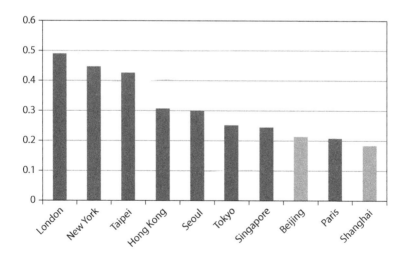

Figure 7.7 Per capita subway mileage in global cities (km/10,000 people). *Source:* World Bank.

4. Clean Energy Needs a Big Boost

In Beijing's energy composition, coal consumption is disproportion-ately high, totaling 23 million tons in 2012, or 25 percent of total energy consumption (this ratio fell to 20 percent in 2014).[16] Based on Beijing's energy development plan, coal's share is expected to fall to 16.8 percent by 2015, while the municipality's "2013–2017 Work Plan to Accelerate Reduction in Coal Consumption and Develop Clean Energies"[17] envisioned cutting coal as a share of energy consumption to 10 percent or less by 2017.

Clean energy, on the other hand, accounted for 17.8 percent of Beijing's primary energy consumption in 2012, of which 4.5 percent was renewable energy and 13.3 percent was natural gas. Beijing wanted to increase the proportion of renewable energy to roughly 6.1 percent and natural gas to 24.4 percent by 2015. Although these targets are slightly higher than the national average, they are nonetheless low when compared to average levels in other developed countries (see figure 7.8).

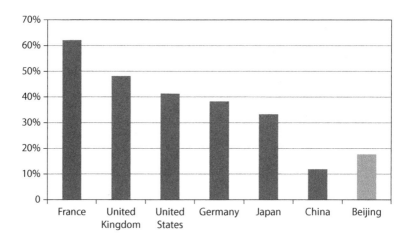

Figure 7.8 Clean energy as a proportion of primary energy consumption, 2011. *Sources: BP Statistical Review of World Energy* (2012); "Beijing Energy Development Plan for the Twelfth Five-Year Plan Period."

5. *Concentration of Heavy Industry*

The share of industry, and particularly heavy industry, in Beijing's economy remains high, which has led to an energy-intensive economy. According to the 2013 *Beijing Statistical Yearbook*, secondary industry accounted for 22.7 percent (18.4 percent industry and 4.3 percent construction) of Beijing's economy, and heavy industry accounted for approximately 85 percent of industrial enterprises above a designated scale.[18] In other words, heavy industry accounted for approximately 15 percent of Beijing's GDP.

In contrast, other major international urban hubs tend to depend heavily on the services industry, while the proportion of manufacturing is very low (see figure 7.9). Compared to these international cities, Beijing holds great potential to dramatically reduce the proportion of industry and manufacturing. In addition, Beijing's energy intensity (energy consumption per 10,000 yuan of GDP) is roughly four to five times higher than that of developed economies such as the United Kingdom and the United States.

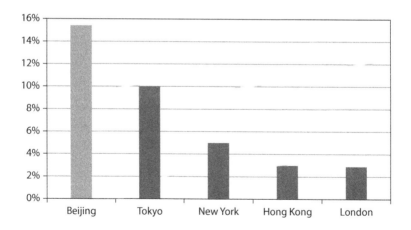

Figure 7.9 Comparison of manufacturing as a proportion of GDP, 2012. *Sources:* *Beijing Statistical Yearbook* (2013); Hong Kong Census and Statistics Department; author's estimates; Jiang Manqi and Xi Qiangmin, "The Status, Role, and Vitality of Manufacturing in the Development of Major International Cities," *Nankai Journal,* 2012.

IV. Regional Compensation and Structural Adjustments

Based on our PM2.5 control model simulations, we concluded that Beijing must establish a new regional coordination mechanism and incentivize significant structural adjustments to meet its PM2.5 target. What follows are detailed discussions of our recommendations in both creating such a mechanism and executing structural changes in the local economy.

1. Creating a Regional Compensation Mechanism for Beijing and Hebei

Because emissions from peripheral areas are the single largest source of PM2.5 emissions in Beijing (accounting for 40 percent), the most difficult—but also the most important—task is the establishment of a regional mechanism for joint emissions reduction. Establishing such a mechanism, in which Beijing pays part of the costs for peripheral areas, can lower the cost of addressing PM2.5 pollution in both the capital and surrounding areas. If this mechanism can be implemented successfully, we estimated that the peripheral contribution may help Beijing cut its PM2.5 level to around 16.8 μg/m³ by 2028, a reduction that is 4.1 μg/m³ greater than would otherwise be achieved with only the existing policies.[19]

It is clear that air pollution in one region affects surrounding areas. If there is no incentive mechanism for this region to account for these externalities, or if the region lacks capacity, its local emissions will be higher than the level needed to maximize the social welfare of the entire region. Also, if administrative means are used to force each region to reduce emissions by the same amount without considering the externality of pollution and the enormous regional differences in the cost of emissions reduction, then it will end up costing more overall.

Therefore, improving the air quality of Beijing's neighbors, such as Hebei, will be ineffective if provinces have to bear the costs entirely on their own. Because Beijing is wealthier and has an urgent need to reduce air pollution, it will be more effective and economical for the Chinese capital to assist regional local governments in reducing their pollution. If the Beijing government reforms industry and subsidizes emissions reduction equipment for its neighbors—Hebei in particular—those steps would minimize the deterioration of Hebei's fiscal condition, economic growth, and employment, while enabling Beijing to reduce its own emissions at lower cost.

I. REGIONAL MECHANISM IN THEORY AND IN PRACTICE

Classic economic theory explains that when firms in one region pollute, they pose not only a local threat, but also generate negative externalities to other regions. For example, the negative externalities can take the form of causing unwanted air pollution, which can have second-order effects of damaging health, increasing healthcare costs, and hurting productivity. Failing to account for such externalities, firms in Hebei will continue to generate high levels of pollution that affect Beijing's air quality. To control pollution in Hebei, it is necessary to establish mechanisms for Hebei to internalize the external costs of emissions. Such a mechanism should enable the province to consider itself and Beijing as an integrated whole when making decisions that have environmental consequences.

Theoretically, there are two approaches that would allow Hebei to internalize this externality. The first is negative incentives: impose hefty taxes on heavy industry in Hebei, or close and/or suspend the operations of heavy industry firms, or require them to install costly emissions reduction technology. Although such measures could, in principle, achieve emissions reductions, they come with considerable downsides such as potentially collapsing the local economy, triggering unemployment, and possibly even catalyzing social unrest. If Hebei lacks sufficient resources and capacity to create new jobs, negative incentives that could lead to an economic recession will be met with fierce resistance.

The second option is positive incentives: let Beijing subsidize Hebei to help its firms deploy emissions reduction facilities and technologies and retrofit coal power plants to natural gas, among other measures. These subsidies essentially serve as "rewards" for Hebei firms to cut emissions, which should make them more inclined to intensify efforts and achieve the effect of internalizing their negative externality. This approach will allow many Hebei firms to continue operating and avert large-scale unemployment as well. Moreover, this will lower Beijing's emissions reduction costs, because investing in Hebei's efforts can offset a portion of Beijing's more expensive local mitigation efforts.

In fact, China has had some success with a regional compensation scheme for dealing with water pollution. Beijing should consider this example when working with its neighbors to devise a similar mechanism for air pollution (see case study box).

Case study: The Xinanjiang Watershed environmental compensation mechanism pilot program

At 260 km, the Xinanjiang River flows through multiple provinces, starting in Xiuning County, Huangshan Prefecture in Anhui province and ending in Zhejiang province at Jiekou Town in He County. The watershed area covers 11,660 m², more than half of which (60 percent) lies within the borders of Anhui province, containing 77 percent of the total watershed population. The Xinanjiang River supplies nearly 70 percent of the water in Qiandao Lake, which is an important source of fresh water for various cities in Zhejiang and is a strategic reservoir for the YRD region. The lake has a surface area of 575 km², normal storage elevation of 108 m, and total storage capacity of 21.63 bcm. Its ecosystem serves multiple functions: water supply, flood prevention, tourism, and biodiversity protection. There is a notable wealth gap between those who live in the headwater and catchment regions of the Xinanjiang River—income levels in upstream Anhui are significantly lower than those in downstream Zhejiang.

For many years, contamination of upstream wastewater in the Xinanjiang River affected the water quality of the downstream Qiandao Lake, and the situation was gradually deteriorating. Although Qiandao Lake is a renowned scenic spot and an important source of drinking water for Zhejiang, systematic measures to manage water quality and control pollution have been consistently absent. The government of Chun'an County, where Qiandao Lake is located, announced ten water pollution control projects, including wastewater and solid waste management in lakeside towns and villages, regulation of aquaculture, leisure vessel waste management, illegal fishing monitoring, and dredging regulations. However, these measures were unable to meaningfully improve water quality in Qiandao Lake. This is because the majority of the pollution in the Qiandao Lake watershed comes from the upstream Anhui province. Control measures in Zhejiang province alone were unable to change polluting behaviors upstream. To deal with the root cause of the pollution, upstream and downstream residents needed to tackle the problem together.

The problem did not lend itself to an easy solution. That is, while upstream Anhui is where the pollution had to be controlled, the poor province did not have the financial resources and incentives to do so. Meanwhile, downstream Zhejiang, a relatively prosperous province, had the financial capacity to address the pollution but lacked the means to mitigate pollution at its origin. Despite the odds, in 2005, a member of the Zhejiang People's Congress proposed the establishment of a regional compensation mechanism to protect the Xinanjiang River watershed. Two years later, the NDRC, Ministry of Finance (MoF), and SEPA (now the MEP) jointly designated the Xinanjiang River watershed the first interprovincial ecological compensation pilot area.

In August 2009, MEP published a draft plan for this pilot program. In October 2010, MoF and MEP jointly provided initial capital of 50 million yuan ($8 million). In March 2011, the ecological compensation pilot was formally launched, and 200 million yuan ($33 million) was allocated for use in the headwater region for pollution control efforts and ecological recovery. The ecological compensation plan was formalized in October 2011, after multiple rounds of negotiations that MoF coordinated. Based on this plan, the central government has allocated 300 million yuan ($50 million) each year to Anhui province since 2012 for environmental protection in the headwater region.

In addition, Anhui and Zhejiang reached an agreement: the average values of four water quality indicators (permanganate, ammonia nitrogen, total nitrogen, and total phosphorus) for a three-year period from 2008 to 2010 would serve as the benchmark water quality index (BWQI) for the ecological compensation program. Starting in 2015, if water quality indicators improved above the BWQI at the provincial border (Jiekou Town), then Zhejiang will compensate Anhui 100 million yuan ($16 million). If not, then Anhui will in turn compensate Zhejiang the same amount. If water quality indicators remain unchanged, then neither party pays.

To enforce the pilot program, the government of Huangshan prefecture established leadership groups and a bureau of watershed construction protection. Pollution prevention and control and ecological recovery plans for the headwater region and Qiandao Lake are currently being formulated. The government of Huangshan prefecture is leading pollution prevention and control while MoF heads ecological recovery. The Huangshan plan includes a series of projects targeting urban point source and nonpoint source pollution, intensifying efforts to monitor and publicize industrial pollution, and increasing the level of control. This plan requires a total investment of 40 billion yuan ($6.5 billion) in over 500 subprojects over five years. In 2011, 350 million yuan ($60 million) was invested in sixty-one subprojects under the ecological compensation program. This investment covers many areas, including nonpoint source pollution controls in agriculture and rural villages, urban waste management, and industrial pollution controls.

This plan in action is the equivalent of Zhejiang province redistributing funds to Anhui to undertake pollution control in the watershed. At the same time, in the event of failure to meet targets, the mechanism of reverse payment effectively acts as a check on ensuring local government compliance and implementation. For Zhejiang, having the control efforts conducted in Anhui not only saves money compared to the cost of undertaking such efforts in Zhejiang itself, it also improves efficiency. For Anhui, receiving funding for environmental protection gives it the ability and capacity to manage its polluting firms. Therefore, this mechanism is mutually beneficial.

Sources: He Haifeng (2012), He Cong (2014).[20]

II. HOW TO MAKE A REGIONAL MECHANISM WORK FOR JING-JIN-JI Rather than three separate schemes, a single, unified mechanism for emissions reduction should be applied to the Jing-Jin-Ji region. This combined effort means that emissions reduction actions taken in each province and municipality will affect the results in the others. For example, if Beijing moves heavy industry and thermal power plants to Hebei, it would appear to reduce local coal consumption but would increase emissions in Hebei. And because Hebei emissions will end up blowing back into Beijing, the latter would still not be able to achieve its emissions reduction target.

The objective of the "Detailed Implementation Plan to Prevent and Control Air Pollution in Jing-Jin-Ji and Peripheral Areas" is to reduce the PM2.5 level for the entire region by 25 percent from 2012 to 2017.[21] We believe that a unified scheme should be adopted to implement measures at the lowest cost for the region as a whole. This does not mean that each local region will need to achieve 25 percent reductions. For example, considering the different costs for emissions reductions in the three areas, a hypothetical scheme could be 40 percent reduction in local emissions in Hebei, 25 percent in Tianjin, and 15 percent in Beijing. The intensity of reduction efforts should be the greatest in the locality where costs are lowest, and conversely, the extent of reductions can be lowest in the area where costs are highest. In this fashion, the emissions reduction target for the entire region can be achieved at a lower cost than would be possible under a scheme that requires each locality to cut 25 percent.

The regional compensation mechanism can facilitate cost sharing among the three regions so that all benefit. Table 7.3 presents a hypothetical comparison of emissions reduction costs under the two systems, based on a number of assumptions. It demonstrates that the compensation mechanism can lower emissions reduction costs while achieving the same results. Alternatively, given a fixed amount of total investment in emissions reduction by the three localities, greater reduction results can be achieved using the compensation mechanism.

The key to designing an effective regional compensation mechanism lies in quantifying the costs of the various measures used in the

TABLE 7.3

Comparison of Emissions Reduction Costs Under Two Hypothetical Schemes (in yuan)

	Acting alone	Using a regional compensation mechanism	
	Expenditure (= contribution)	Amount spent locally	Amount contributed
Municipality B	1 trillion	650 billion	800 billion
Municipality T	500 billion	350 billion	400 billion
Province H	800 billion	950 billion (of which 150 billion is compensation from municipality B and 50 billion is compensation from municipality T)	750 billion
Total cost	2.3 trillion	1.95 trillion	1.95 trillion

Source: Author's estimates.
Note: All figures are based on hypothetical scenario.

different areas to reduce local emissions—for instance, the cost per ton of direct PM2.5 emissions reduced. It is also important to propose a set of emissions reduction actions that have the lowest total cost but that can still achieve the target for the entire region—in this instance, a 25 percent reduction. This combination of measures can be a matrix comprising of how much money needs to be invested by Hebei, Beijing, and Tianjin in the respective areas of power plant and industrial desulfurization and denitration, boiler modification, natural gas and renewable power generation subsidies, and alternative energy vehicle subsidies. Since Hebei's investment needs will significantly exceed its financial capacity, Beijing and Tianjin should step in to cover Hebei's shortfall.

One of the primary reasons that subsidizing Hebei will lead to lower costs for the region overall is because there is still great potential in Hebei to apply low-cost emissions reduction technologies. That is, existing emissions reduction equipment and their operation in Hebei lag far behind those in Beijing. As of the end of the eleventh FYP period in 2010, Beijing had already achieved elimination of small thermal power generators of 100,000 kW or less, completion of exhaust desulfurization and denitration projects at major power plants, and basic completion of the large coal-burning boiler exhaust desulfurization projects. Therefore, for Beijing, there is limited potential for emissions reduction in power generation and industry because much of the "low-hanging fruit" has already been picked.

In contrast, although Hebei has made some progress in desulfurization in the power sector, the potential for emissions reduction in other industries remains high. For example, in Hebei's cement industry, the overall denitration rate is still less than 70 percent, and in the glass industry, the overall desulfurization rate is less than 85 percent and the denitration rate less than 70 percent. According to the "Hebei Atmospheric Pollution Prevention and Control Target Responsibility Statement," in 2014–2015 alone, there was estimated to be more than 7,000 m² of steel sintering facilities, 45,000 tons/day of cement output capacity, over 15 GW of electricity requiring denitration, as well as 5.7 million tons of petrochemical capacity requiring desulfurization, among others. Moreover, the steel and power sectors still need dust removal retrofits, while the petrochemical industry needs VOC controls.

Installing more desulfurization equipment in Hebei is a relatively cheap way to cut emissions directly. For instance, the annual cost of PM2.5 emissions reduction, based on equipment depreciation for desulfurization, is only around 800–1,200 yuan ($130–$194)/ton. However, if Beijing were to directly subsidize more natural gas power generation, the cost per ton of PM2.5 reduction is five to ten times that of desulfurization and denitration efforts in Hebei. Moreover, once Beijing sufficiently reduces local coal consumption, the capital city will need to rely primarily on controlling automobile emissions to continue to make progress on reductions. This would likely require extensive subsidies for clean energy vehicles, which would cost up to hundreds of times more per ton of PM2.5 reductions than implementing desulfurization and denitration in Hebei.

If it becomes too difficult to establish a compensation scheme for the Jing-Jin-Ji region for a variety of political and economic reasons, then regional governments should consider launching an alternative compensation mechanism between Beijing and Hebei only, with simple objectives. Both governments could first arrange for experts to propose a set of emissions reduction projects in Hebei that have the highest cost-benefit ratio—defined as the best emissions reduction results per unit of investment by Beijing. To put it more colloquially, these are projects that have the biggest bang for the buck when it comes to Beijing's investment. These projects will then have to demonstrate that they result in greater PM2.5 emissions reduction than would be otherwise achieved if Beijing invested the same amount in projects within its own municipality.

Funding for this compensation scheme could come from Beijing's fiscal coffers, or from electricity price subsidies paid by Beijing residents, or from fees paid by firms in Beijing to purchase emissions quotas. To operationalize this compensation mechanism would require selecting a set of carefully vetted lowest-cost emissions reduction projects in Hebei, and based on the rate of benefit to Beijing—for example, 1 ton of emissions reduction in Hebei equals 0.25 ton of emissions reduction in Beijing—the Chinese capital agrees to contribute a proportional amount of money for these projects, in this case 25 percent. A joint working group would be responsible for supervising projects and presenting regular progress reports. If expected emission reduction results are not achieved, then Beijing would stop funding the next stage of the projects.

Beijing should also assist Hebei in acquiring central government funding for environmental efforts and should offer technical assistance. With financial support from the central government and from the Beijing municipal government, Hebei's incentives and capacity to execute air quality control measures will be greatly enhanced. Under the new mechanism, Beijing and Hebei will achieve a more efficient and mutually beneficial arrangement—in theoretical terms, a Pareto improvement for the social welfare of both regions.

Since late 2013, the interregional coordination among Jing-Jin-Ji on air pollution control has been concentrated in technical and administrative areas, such as air quality monitoring and joint efforts to implement coal consumption reduction, closure of polluting manufacturing facilities, and prohibiting vehicles that do not meet minimum emissions standards. So far, very little has been done to set up a regional compensation mechanism for air pollution control.

2. Cap Vehicle Fleet Size, Drive Less, and Build More Subways

In terms of changing Beijing's transportation structure, we propose the following measures:

1. Control the annual growth in the motor vehicle fleet to 2 percent through 2020, and keep it at 1 percent after 2020, so that the total vehicle fleet will be less than 7 million by 2028.
2. Reduce the vehicle utilization rate an additional 10 percent beyond what is assumed in the existing policy scenario.
3. Expand total subway length from the current 660 km to 1,500 km, and increase subway carrying capacity so that the proportion of subway travel in Beijing's central areas will increase from the current 16 percent to 40 percent or above by 2028.

These policy proposals, if faithfully implemented, should lead to flat growth in terms of road transportation volume. But because subway transportation volume growth will average 10 percent a year, this will allow total transportation volume growth to average 3.5 percent a year. Reducing growth in road transportation volume, improving traffic congestion, and cutting the amount of driving time per unit of mileage will all contribute to reducing vehicle-related emissions. Relative to the "planned environmental measures" scenario, these adjustments in Beijing's transportation structure will reduce the PM2.5 level by 5.1 $\mu g/m^3$ through 2028. Specific measures to manage vehicle fleet growth and boost subway travel are discussed below.

I. KEEP VEHICLE FLEET GROWTH TO UNDER 2 PERCENT At the end of 2013, Beijing had 5.42 million motor vehicles—the highest number among Chinese cities—and 80 percent of these vehicles were private cars. The massive fleet size has resulted in awful traffic congestion and air pollution. Policies such as the auto lottery system have had an impact on slowing motor vehicle growth to 4 percent in 2013, down from the 11 percent average growth from 2007 to 2012. But our projections suggest that to achieve the PM2.5 target for 2028, it will be necessary to further cut the growth of the vehicle fleet to an average of 2 percent a year through 2020 and less than 1 percent after 2020.

II. CHANGE CAR LOTTERY SYSTEM TO A LICENSE PLATE AUCTION SYSTEM In 2011, the Beijing Municipal Commission of Transport issued the "Beijing Municipality Provisional Regulations to Adjust and Control the Number of Small Passenger Vehicles," which rolled out a lottery system for license plates in a bid to contain auto growth. In the three years after its implementation, the lottery system has slowed the growth of the total number of motor vehicles; at the same time, it has become increasingly difficult for an individual to win the lottery. By early 2014, the probability of winning the lottery was less than one in a hundred,[22] making it nearly impossible for individuals who can afford to buy a car to obtain a license plate. At this rate, a Beijing resident would need to, on average, participate in the lottery one hundred times—or continuously for about eight years, since the lottery is run once a month—in order to win once.

Shanghai was China's first local government to institute a license plate auction system, in 1986. To further ease Shanghai's traffic congestion, an auction system for additional passenger vehicle quotas was instituted in 1994. The auction system certainly played a part in limiting the growth of private vehicles in Shanghai, which totaled 1.83 million vehicles in 2014,[23] only 42 percent the size of Beijing's fleet (4.37 million private vehicles in 2014).[24] Recently, Tianjin and Guangzhou implemented similar license plate auction systems.

Compared to a lottery system, three advantages of an auction system stand out, in our view. First, the lottery system does not allocate license plates to those most willing to pay, resulting in an unnecessary loss of utility to consumers as a whole. To put it more bluntly, under the lottery system, the government essentially fixes the price of highly valued license plates at zero, resulting in an artificial increase in demand. This leads to a perverse situation in which those who truly need a plate find it nearly impossible to obtain one.

The following example illustrates the loss of utility. The lottery system offers all participants the same probability of obtaining a license plate, irrespective of the marginal utility afforded them. In other words, the people who have an opportunity to win the lottery are generally not those most in need of a license plate (defined as the people having the highest marginal utility). For example, assume that of the million people who participate in each lottery, 20,000 win a plate. Furthermore, assume a linear distribution of the marginal utility of the license plates to these people from virtually zero to approximately 100,000 yuan ($16,000), which is the approximate price of a license plate in a typical Shanghai auction. Finally, assume that winners are randomly selected among the participants, then the average utility of a lottery number per winner is 50,000 yuan ($8,000).

Thus, under the lottery system, the marginal utility of each person who wins a license plate is on average 50,000 yuan ($8,000) less than the marginal utility of a license plate obtained under the auction system. The total utility loss for the hypothetical auction with 20,000 license plates won is therefore 1 billion yuan ($161 million), and the total utility loss for the twelve auctions held each year is 12 billion yuan ($2 billion).

Second, like Shanghai, Beijing can use the additional fiscal revenue generated from the license auction system to invest in projects ranging from air pollution control technologies and subway construction to deployment of clean energy. By not using an auction system, Beijing is essentially forgoing this important source of fiscal revenue.

Third, putting an auction system in place will cause a substantial decline in license plate demand in Beijing. For example, 1,818,640 people participated in the Beijing lottery of February 2014,[25] nearly forty times the 45,758 bidders who participated in the Shanghai auction in the same period.[26] Relying on an auction system will raise the price of a plate and thus reduce demand. It will increase the success rate in obtaining plates and prevent the waste of public resources on the monthly lotteries in which nearly 2 million people participate but with extremely low probability of winning.

Thus, Beijing should consider implementing a license plate auction system under the following terms:

- Limit all new license plates to a valid term of ten years, and require holders of expired license plates to enter another bidding round to obtain a new license.
- Limit the number of new licenses to be auctioned off so that the net increase in the outstanding number of licenses grow at an annual rate of 2 percent.
- Adhere to the "polluter pays" principle by collecting an annual fee from car owners who received their license plates before the new system took effect.
- Require all public vehicles (except for ambulances, police cars, and fire engines) to participate in auctions, including central government agency and military vehicles with Beijing plates.

III. LEVY CONGESTION FEES AND RAISE PARKING FEES The auction system should be the primary tool used to cut the growth of Beijing's vehicle fleet, while congestion fees and increased parking fees should curb private car usage (the mileage driven per vehicle per year).

First, Beijing should institute road congestion fees. The fee should be collected from private car owners in the city's central areas. Because a congestion fee will not be levied on public transportation, including buses and subways, Beijing residents will be more likely to choose public transportation as driving becomes relatively more expensive. The districts of Xicheng, Dongcheng, Chaoyang, and Haidian should all be considered "central areas" where a congestion fee would apply.

Second, Beijing should consider increasing parking rates for all major parking lots in the city. While the government determines parking rates, property management companies collect the fees. These fees, plus sales tax and road use tax, are paid into the national treasury. Increasing parking rates will lower vehicle utilization rates by making vehicle usage more costly.

IV. BOOST SUBWAY RIDERSHIP Subways are a relatively low-emissions mode of transportation and should be further encouraged in Beijing as the primary mode of urban travel. Beijing should strive to increase subway travel as a proportion of resident travel in central areas from the current 17 percent to about 40 percent in 2028.

However, even if the total subway mileage reaches 1,200 km in 2028, our estimates indicate that it will still be difficult to achieve the PM2.5

target. We recommend that the Beijing subway be lengthened to a total of 1,500 km to reach a per capita mileage of 0.5 km/10,000 people.

Currently, subway financing in Beijing is essentially the responsibility of the municipal government, with most of the funding coming from special project financing. Each year, the municipal government allocates 10 billion yuan ($1.6 billion) in special funds for rail construction, which is shared according to the principle of municipality-district sharing: 80 percent from Beijing and 20 percent from the districts through which the rail passes.

In addition to allocating budgetary funds, the Beijing Infrastructure Investment Co., Ltd. (BII) is mainly responsible for raising funds through debt and equity financing for subway construction. Since it first introduced the concept of public private partnership (PPP) in financing the building of Line 4 in 1994, BII has tried different financing models (see figure 7.10). BII has used the PPP model and the build-transfer model for investing in Lines 4 and 8, respectively. The private capital involved in this financing has increased the rate of subway investment and construction, while introducing operating and management experience from overseas.

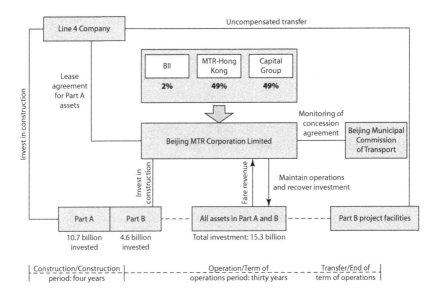

Figure 7.10 Beijing subway Line 4 project operation model (in yuan). *Source:* BII Co., Ltd.

The PPP model uses some government funding and also attracts private capital (e.g., insurance, pension funds, private equity, foreign investment, and businesses) into public infrastructure, with shared risks and returns. Since rail transportation is considered a quasi-public good, the PPP model can separate the public benefits of the project from profitability. The government is involved to ensure public benefit and private investors are incentivized because of reasonable returns. Mechanisms are established to manage risks, and to ensure sustainability of revenue streams, safety, and public benefits.

BII's success with the PPP model should be applied to other subway projects to raise funds for subway construction, allow greater market participation in public infrastructure, and accelerate the expansion of total subway length.

V. INCREASE SUBWAY CARRYING CAPACITY While Beijing's subway carries an enormous number of passengers daily, its operating efficiency is nonetheless relatively poor. As of 2012, the Tokyo subway system totaled 312.6 km with an average daily passenger volume of 11 million.[27] Beijing's subway system, on the other hand, is actually longer, at 465 km, but carries fewer people, with average daily volume of 10 million. This results in a carrying capacity (passengers carried per km) that is just 60 percent of the carrying capacity of the Tokyo subway system, suggesting that the Beijing subway's carrying capacity can be bolstered by about 20 percent. We believe this can be achieved with various means.

First, the model of train cars used on existing lines should be changed from B-size to A-size. Initially built for strategic purposes, the Beijing subway used B-size cars with a smaller carrying capacity for Lines 1 and 2 rather than the larger A-size cars. The maximum carrying capacity for a single B-size car is 240 passengers, compared to 310 passengers for A-size cars.

Second, the length of subway trains should be increased in some areas to boost total carrying capacity of each train. Given that twelve new subway and light-rail lines are being built or planned in Beijing,[28] the need for longer trains should be considered in the early planning stage. Larger train cars and longer trains (with more cars) will require longer train platforms.

Third, subway system planning can be improved to reduce passenger traffic bottlenecks and increase total carrying efficiency. Traffic control, increased line density, optimized and modified channels, and staggered peak hours may improve operating efficiency. For example, currently only a single vertical line connects south to north of the ring roads. This line becomes very stressed during morning and evening rush hours, significantly limiting carrying capacity. Beijing should consider building several more vertical lines to alleviate the stress.

VI. USE NEW STREAMS OF REVENUE TO BOOST PUBLIC TRANSPORTATION FUNDING Beijing should adopt an approach similar to Shanghai's by using revenue from new fees to fund the adjustments needed in the transportation sector. Based on the Shanghai experience, fiscal revenue from license plate auctions and congestion fees could approach 10 billion yuan ($1.6 billion) annually. But if the municipal government isn't transparent or fails to communicate to the public how these new fees will alleviate traffic congestion, reduce smog, and help fund public infrastructure, these policies will be met with public resistance and political pushback. Therefore, to win and sustain public support for these actions, the Beijing government could consider earmarking this additional revenue for public transportation and environmental projects and disclose its usage.

Beijing could also use some of the extra revenue to finance emissions reduction efforts in Hebei, such as deploying clean energy, extending subway lines, upgrading train stations, and procuring natural gas-powered buses. Some of the money could also be used to phase out vehicles that fail to meet new emissions standards.

3. *Dramatically Bolster Clean Energy*

From 2013 to 2028, Beijing should boost clean energy as a proportion of its primary energy consumption from the current 18 percent to 65 percent and simultaneously reduce the proportion of coal from 25 percent to 5 percent. This dramatic adjustment to Beijing's energy composition should satisfy the energy needs of continued, though more moderate, economic growth, and also lead to further reduction of the PM2.5 level by about 1.9 $\mu g/m^3$ compared to the scenario of relying on environmental measures alone.

Altering the energy structure will entail replacing coal with more natural gas and renewables, as well as boosting the penetration of alternative energy.

I. INCREASE THE PROPORTION OF NATURAL GAS IN ENERGY MIX Natural gas currently provides about 14 percent of Beijing's primary energy needs, and the municipality intends to raise that to 24 percent by 2015, according to its twelfth FYP on energy development. But Beijing has relatively abundant natural gas resources, receiving supplies via the pipeline from Russia and from domestic suppliers because, as the national capital, it is considered a priority market.

As such, Beijing should consider raising natural gas to 50 percent of its primary energy consumption by 2028 and harness it for various end uses. For one, Beijing should strongly promote natural gas buses and taxis, as well as gas boilers. Two, the city should continue to improve gas pipeline infrastructure within Beijing and accelerate fuel switching from coal to gas as the primary source of residential energy consumption. Moreover, the municipal government should actively incentivize natural gas use in surrounding rural villages and initiate a pilot program to build gas pipelines in those areas. Finally, Beijing should expand natural gas storage, filling, and distribution facilities, as well as encourage more gas consumption in industrial production, refrigeration, motorboats, and automobiles.

II. RAISE THE PROPORTION OF RENEWABLE ENERGY Renewable energy is expected to rise from 4.5 percent of Beijing's primary energy consumption to 6.1 percent by 2015, according to the twelfth FYP on energy development. This proportion should be further increased to 15 percent by 2028, with priorities placed on expanding the use of solar, geothermal, and biomass resources.

More specifically, solar heating and geothermal heat pump technology should be more widely used for home heating, refrigeration, and residential water heating. Finally, alternative energy vehicles and corresponding infrastructure, such as charging stations, should be more aggressively deployed.

III. FOLLOW SPECIFIC POLICY RECOMMENDATIONS First, subsidies for clean energy should be raised significantly. At present, national subsidies for clean energy as a proportion of GDP are a meager 0.2 percent, and additional subsidies from the Beijing government are roughly the same. To drastically boost clean energy consumption, the government needs to increase on-grid power tariff subsidies for renewable energy and offer subsidies for maintaining clean energy vehicles and equipment. The main sources of these subsidies could come from user fees, such as by increasing electricity tariffs and pollution fees.

Second, the government should purchase clean energy vehicles through government procurement. Public transportation, environmental sanitation, and government agencies should lead the way by requiring that 80 percent of newly purchased or modified public transportation vehicles be clean energy vehicles. At the same time, low-speed EVs, such as three-wheeled motor vehicles and low-speed trucks, and natural gas vehicles should be used more widely, and the government should encourage development of natural gas filling stations.

Third, the resource tax on coal should be raised and a carbon tax levied to support the government's funding needs. Beijing should also consider significantly raising fees on SO_2 and NOx emissions.

4. Reduce Heavy Industry in the Local Economy

Industry currently accounts for about 18 percent of Beijing's GDP. As part of the broader secondary industry, heavy industry makes up 85 percent of the output of industrial firms above a designated scale.[29] This is in stark contrast to national capitals in developed countries where heavy industry is essentially absent. Based on our model, we recommend that Beijing reduce the proportion of heavy industry to 10 percent of GDP or less by 2028. Compared to relying on environmental measures only, the structural adjustment to industry can reduce Beijing's PM2.5 level by about 5.3 µg/m³. The following specific steps and policies should be pursued:

- Do not approve any new heavy industry or chemical projects, and move a large portion of existing firms in those sectors out of Beijing. This way, through the attrition of old production capacity, the proportion of heavy industry will gradually decline.
- Encourage and promote the transition of heavy industry toward high tech and services industries. Firms that have transitioned successfully globally, such as General Electric, which diversified from a chemical and electric company into biopharmaceuticals, renewable energy, and high-end manufacturing, have done so

by strengthening R&D and pursuing mergers and acquisitions (M&As) to achieve product upgrades and shift into services. Drawing on these global experiences, the Beijing government should focus on helping heavy industry firms shift from expanding production capacity to cleaner production via R&D and M&As.

• Accelerate the replacement of coal and oil with renewable energy.

5. *Breakdown of Contributions to Beijing's PM2.5 Reduction*

To reiterate, our model simulation indicates that if all the existing and planned environmental policies are implemented and all the structural adjustment targets are realized, Beijing will be able to reduce its PM2.5 level to 35 µg/m³ by 2028 (see fig. 7.11).

Generally speaking, of the 55 µg/m³ reduction needed for Beijing to hit its PM2.5 target, existing and planned environmental measures and regional efforts contribute about 70 percent (38 µg/m³). The remaining 17 µg/m³ of reductions needed must come from the structural adjustment actions we recommended above (see table 7.4).

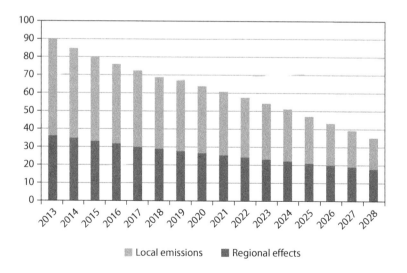

Figure 7.11 PM2.5 reductions under structural adjustment scenario (µg/m³).
Source: Author's estimates.

TABLE 7.4
Contribution to Emissions Reduction Under Structural Adjustment

	Extent of emissions reduction ($\mu g/m^3$)	Contribution to reductions (%)
Existing environmental measures	23.8	43.3
Structural adjustments	13.1	23.8
1) Transportation	5.1	9.3
2) Energy consumption	1.9	3.5
3) Industry	6.1	11.1
Peripheral areas' environmental actions	14.1	25.6
Regional compensation mechanism	4.1	7.5
Total reductions	55	100

Source: Author's estimates based on PM2.5 control model.

Note: There is a direct relationship between the emissions reduction results above and the sequence of implementing the reduction measures. If the structural adjustment measures come before existing environmental actions in the simulation, the results of the structural adjustments would be substantially greater due to the initial lower base.

CHAPTER EIGHT

How to Deal with Coal

Introduction

Coal has long held a dominant position in China's overall energy consumption. In 2013, coal accounted for 71.6 percent of total energy output and 59.3 percent of total energy consumption in China.[1] According to BP, China consumed 50.6 percent of the world's coal in 2014, compared to only 29 percent in 2000.[2]

A coal-dominated energy structure and the rapid growth of coal consumption have generated a number of serious problems, environmental pollution being the most prominent. Since January 2013, many Chinese cities have suffered severe PM2.5 pollution, alarming both the Chinese public and government. About half of the PM2.5 pollution comes from emissions related to coal burning. Therefore, reducing coal consumption needs to be a key focus in any strategy for bringing PM2.5 emissions under control.

Efforts to address the coal problem can be placed in two categories: (1) reducing emissions per unit of coal consumption and (2) reducing the absolute level of coal consumption.[3] The solution to the first problem involves wider deployment of clean coal technologies. But as shown in previous chapters, even if clean coal technologies are applied to the fullest extent, it is still not enough to meet the 2030 PM2.5 target. Therefore, it is also necessary to take action on the second problem to achieve the target. In the following sections, we propose a series of economic and fiscal policies, based on the results of our PM2.5 model simulations, to significantly slash coal consumption.

Brief Literature Review on Coal Consumption

Before diving into an extensive analysis and discussion of how to reduce coal consumption, it is worth reviewing some of the existing literature and studies on this matter (see table 8.1).

A substantial body of existing work uses econometric models to study the demand elasticity of coal, including Masih (1996),[4] Kong Xianli (2010),[5] and Zhang Lei (2013).[6] While these studies reflect how coal consumption is affected by factors such as price, economic growth, and industry structure, they cannot be easily adapted to assess the effectiveness of policies to curb coal consumption. Their shortcomings include: (1) in some of the studies, the estimated elasticity is quite unstable because it is difficult to control the effects of other variables; (2) they lack a microeconomic foundation, in particular, a framework for general equilibrium; and (3) in these models, it is difficult or impossible to simulate the combined effects of multiple policies that are implemented simultaneously.

As a policy analysis tool that is based on the general equilibrium theoretical framework, CGE models are better suited to simulating the effects that fiscal and other policies have on energy consumption and energy composition. Over the past several decades, CGE models have been widely applied in research on policy impacts such as taxes, international trade, income distribution, and the environment and energy. In terms of coal-related research in China, Wei Weixian (2009),[7] Xu Xiaoliang (2010),[8] Guo Ju'e (2011),[9] Lin Boqiang (2012),[10] and Yin Aizhen (2013)[11] all used energy/resource CGE models to simulate the effects of resource taxes and other policies.

Wei's work (2009), for example, simulated the effects of hiking resource taxes, reducing export tax rebates for heavy industry, and adjusting the tertiary industry structure on China's GDP, energy intensity, and household welfare, respectively. Wei concluded that to be effective, resource taxes must be based on a rational and transparent energy pricing mechanism, and that cutting the proportion of heavy industry and raising the proportion of tertiary industry can effectively reduce energy intensity. However, Wei's analysis did not address internal adjustments to the energy composition or ways to increase the proportion of tertiary industry.

Many Chinese economists have attempted to determine the impact of various levels of resource tax on different aspects of the economy. For instance, Xu (2010)[12] employed a resource CGE model to focus more narrowly on the impact of the resource tax on coal at three different rates: 4 percent, 6 percent, and 8 percent. Xu found that revenue is highest and social welfare improvement the greatest when the tax rate is set at 6 percent, while resource utilization efficiency is the highest when the rate is set at 8 percent. Guo (2011)[13] also targeted the resource tax but wanted to determine its impact on coal demand. He used an energy CGE model to simulate changes in prices and demand for coal when the tax is set at rates of 5 percent, 10 percent, 15 percent, 20 percent, 25 percent, and 30 percent. Guo found that coal demand would decline by 2.25 percent, 4.1 percent, 5.64 percent, 6.94 percent, 8.05 percent, and 9.03 percent, respectively, at each of those tax rates.

Lin (2012) used a dynamic CGE model to analyze the macroeconomic effects of a price-based (ad valorem) coal resource tax. If ad valorem tax rates were set at 5 percent, 7 percent, and 12 percent, Lin found that the effects on China's GDP would be −0.153 percent, −0.223 percent, and −0.377 percent, respectively, while energy intensity would decline by 0.134 percent, 0.171 percent, and 0.238 percent, respectively. Finally, Yin (2013) used a CGE model to simulate the effects of resource taxes set at 2 percent, 4 percent, 6 percent, and 8 percent on output and prices in the coal and oil/gas sectors. Yin's results suggest that coal output would decline by 2.6 percent, 5.8 percent, 7.5 percent, and 9.3 percent, respectively, while oil and natural gas output would also decline significantly.

The works cited previously serve as important references for this study, but they also fall short in certain respects when it comes to analyzing China's environmental pollution issues. First, there are enormous differences in PM2.5 emissions generated by different energy sources and the different ways in which they are used. For example, PM2.5 emissions resulting from coal consumption are far greater than those from natural gas and coal power using IGCC technology. Although previous research has noted the need to raise resource taxes to improve energy utilization efficiency, it has not considered ways to increase subsidies for clean energy to improve the energy composition and reduce emissions. Second, the policies proposed in the existing literature primarily center on raising coal prices via taxes to lower emissions, but do not necessarily consider incentives for firms to boost adoption of emissions reduction equipment. Third, many coal producers are already saddled with a heavy tax burden, and their survival is in jeopardy. So an approach largely consisting of tax increases that may bankrupt firms could result in massive unemployment and social unrest. This is another issue to which the existing literature has provided few answers.

Mindful of existing shortfalls in the literature, we have built an energy CGE model to simulate a more complete set of economic and fiscal policies to control coal consumption and improve the energy composition. Such policies include increasing the coal resource tax rate, instituting a carbon tax, subsidizing clean energy, and incentivizing firms to increase their utilization of emissions reduction equipment via pollution fees.

I. How Much Should Coal Consumption Fall?

For the purposes of this analysis, "coal consumption" is defined as the consumption of conventional coal, which excludes coal used for coal gasification power generation, coal-to-gas, and coal-to-oil but includes coal used in thermal power generation and by industries and households. We use the PM2.5 control model to simulate China's pollution levels under three scenarios:

- **No reform**—current energy composition and emissions intensity continue into the future.
- **Partial reform**—energy composition remains essentially unchanged, but end-of-pipe pollution measures are applied and transportation structure is improved.
- **Full reform**—in addition to the measures under the partial reform scenario, the energy structure is substantially altered by reducing the proportion of coal consumption and deploying clean energy.

The model uses the same set of assumptions on GDP growth, energy and transportation elasticity coefficients, and average energy consumption and transportation volume growth that were detailed in part one and in various previous chapters.

TABLE 8.1
Comparison of Existing Research on Coal Consumption

	Research subject	Brief Summary
Econometric approach	Coal demand elasticity	Studies factors—such as price, revenue, energy substitution, industry structure, and transportation costs—that have affected coal demand in China in the past. But the research does not examine ways to control coal consumption or the effectiveness of policies to control coal consumption.
CGE model approach	Effects of resource tax on coal/energy consumption	Quantitatively evaluates the ways in which various resource tax rates affect the Chinese economy and energy composition. However, most of these studies fail to consider the impact of other policies, as well as adjustments to the energy composition and industrial structure on coal consumption.
Our approach	1. Coal resource tax and carbon tax	Evaluates the impact of various tax policies on coal consumption.
	2. Pollution fees	Studies how an increase in pollution levies would drive cost-minimizing firms to install emissions reduction equipment.
	3. Clean energy subsidies	Studies how energy composition may shift toward cleaner energy sources as a result of subsidies.
	4. Industry structural adjustment	Studies how a decline in the proportion of heavy industry in the economy may reduce coal consumption.
	5. CGE evaluation and other quantitative methods	Analyzes the first three policies using a CGE model, and directly estimates the impact of industrial policy using energy coefficients from the input-output table.

Source: Author.

1. PM2.5 Trends Under Various Scenarios

I. No Reform Under this scenario, we expect China's coal consumption to increase by an additional 50 percent from 2013 to 2030, averaging 2.3 percent growth per year,[14] and the number of passenger vehicles to increase by 300 percent, from 100 million to 400 million vehicles during the same period. Assuming emissions per unit of coal consumption and single vehicle emissions remain unchanged, the average PM2.5 concentration in cities will rise 70 percent, reaching a severe level of 110 $\mu g/m^3$ by 2030.

II. Partial Reform This scenario assumes a combination of end-of-pipe control measures and some changes in the industrial and transportation structures that are strictly enforced. These assumptions include:

- Limit passenger vehicle fleet to 250 million (although experts predict 400 million vehicles) by 2030;
- Increase total rail mileage by 60 percent and quadruple total subway mileage from 2013 to 2020; increase rail mileage by an additional 60 percent and subway length 230 percent from 2020 to 2030;
- Cut single vehicle emissions/km by 82 percent through more stringent fuel quality standards and auto emissions standards and through adoption of electric/natural gas vehicles from 2013 to 2030;
- Use clean technologies in the coal industry to reduce PM2.5 emissions per unit of coal consumption by 69 percent from 2013 to 2030, averaging 6 percent annual reductions;
- Reduce secondary industry as a proportion of GDP by 9 percentage points and increase the proportion of urban resident subway travel from 7 percent to 25 percent from 2013 to 2030.

Even if these actions and structural improvements are fully realized, our model shows that if the proportion of coal consumption remains the same, the average PM2.5 level in Chinese cities will still be 38 $\mu g/m^3$ by 2030, higher than the target of 30 $\mu g/m^3$.

III. Full Reform Under this scenario, the findings from our model suggest that, in addition to fully implementing the partial reforms,

China's coal consumption should decline by an average of 0.8 percent a year from 2013 to 2030, for a cumulative reduction of 14 percent.[15] In other words, to achieve the PM2.5 target of 30 μg/m³, coal consumption must decline by a total of 14 percent and coal as a proportion of primary energy consumption needs to fall dramatically from 67 percent to 32 percent.

Of course, this does not mean that coal consumption must decline immediately. In fact, we designed different growth rates for different stages of the 2013–2030 period. If clean coal technologies, such as desulfurization and denitration, are enforced widely and central heating is implemented in the next several years, coal consumption can still grow, albeit at a much slower average rate of about 1 percent from 2013 through 2016, but should decrease substantially starting in 2017. (This forecast also assumes that clean energy will not replace coal on a large scale over the next several years.)

Based on these results, the bottom line is that coal consumption should peak in 2016, about a decade earlier than most forecasts (see more detailed analysis in chapter 3).

2. IMPACT OF FISCAL AND TAX POLICIES ON COAL CONSUMPTION Under the "no reform" scenario, we project that between 2014 and 2017 coal consumption growth will continue a modest deceleration to about 3 percent a year. But to hit the 2030 PM2.5 target, growth has to slow to just 1 percent over the same period, with coal consumption peaking in 2016. In other words, average annual growth of coal consumption should fall by 2 percentage points for a total decline of 8 percentage points over four years.

To achieve such ambitious change in total coal consumption growth in such a short time, robust policy support is needed. While China has relied extensively on administrative means, such as shuttering and suspending coal operations, the economic costs associated with this approach are too great and the results difficult to sustain. Instead, the government should use economic means, consisting primarily of fiscal and tax policies, to incentivize consumers to use less coal and firms to produce less coal, shifting both toward the consumption and production of cleaner energy.

What follows is an attempt to answer the question of what combination of fiscal policies, as well as the intensity of those policies, is needed to meaningfully move China away from what has been a coal-based economy.

Coal Demand Structure and Industry Consumption Since the 1980s, coal has consistently accounted for more than two-thirds of China's energy production and consumption. Coal demand in China primarily comes from power generation, heating, coking, and manufacturing. Of these sectors, power generation and coking have seen the most dramatic growth in coal use, with consumption in power generation rising from 41 percent in 2001 to 51 percent in 2011. Beyond conventional use, coal consumption for the purposes of coal gasification and coal-to-liquids—8.7 million tons and 3.45 million tons of coal, respectively—is just 0.35 percent of total consumption. In 2012, coal accounted for 77 percent of energy production and 67 percent of energy consumption (see figure 8.1).

When it comes to the fuel that powers industry, coal is indisputably king. Although industrial value-added accounts for 40 percent of China's GDP, coal use by industry (including power generation) accounts for 78 percent of all coal consumed (see figure 8.2). Breaking it down further, heavy industry (including sectors such as power generation, oil and gas, steel, nonferrous metal, cement, chemical,

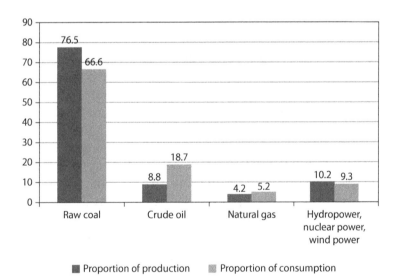

Figure 8.1 China's energy production and consumption composition, 2012 (%). *Source: China Statistical Yearbook* (2012).

Note: Numbers do not add up to 100 because of rounding.

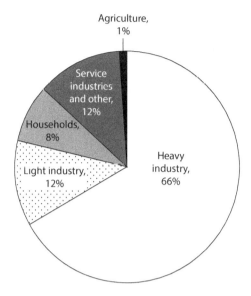

Figure 8.2 Coal consumption by industry, 2012. *Sources:* Author's estimate based on industry data; NBS.
Note: Numbers do not add up to 100 because of rounding.

and machinery) as a whole is just 29 percent of GDP and makes up 70 percent of industrial value-added, but accounts for 66 percent of total coal consumption. The coal consumption intensity of heavy industry is nine times that of the services industry (see figure 8.3).

3. POLICY RECOMMENDATIONS As noted, the Chinese government has, up until now, largely relied on administrative means to curb coal consumption. Although forced shutdowns of coal boilers or pressuring local governments to meet coal consumption reduction targets may produce fast results, monitoring costs are extraordinarily high and the potential for corruption and data falsification is significant. Just as important, the impacts of such administrative actions tend not to last. Moreover, if forced closures become so intense that economic inputs, such as capital and labor, have insufficient time to adjust, serious unemployment and social instability could result.

Finally, reliance on administrative actions also means forgoing a source of fiscal revenue that would otherwise be generated from using economic tools such as raising taxes and fees (see box on previous page).

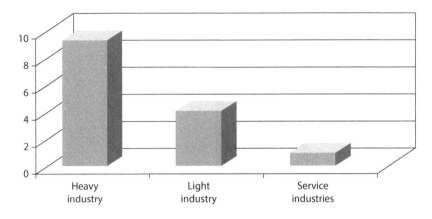

Figure 8.3 Coal consumption intensity across sectors. *Source:* Author's estimates.
Note: Coal consumption per unit of GDP in the services industry is calculated as 1.

Taxes and Fees in the Coal Industry

Many types of taxes and fees, which vary by locality, are already imposed on coal. In terms of the tax burden as a whole, estimates also vary. Peng Yi, chief accountant of the China National Coal Group Corporation, estimated that "in 2011, China National Coal Group's average tax burden on coal for sale was 126 yuan ($20)/ton." Di Deyuan, deputy general manager of the Yitai Group, estimated that "taxes and fees on a ton of coal were 120–130 yuan ($20–$22) in Shanxi, 140 yuan ($24) in Shaanxi, and 133 yuan ($22) in Inner Mongolia." Zhang Bo, department head of the finance department at Shandong Energy Group, estimated that its taxes and fees were about 160–170 yuan ($27–$28)/ton.[16] If the average sales price of coal has been roughly 500 yuan ($85)/ton in recent years, then the overall tax burden is between one-fourth and one-third of the coal price.

The VAT has the greatest impact on coal prices, demand, and production. In 2011, the real VAT burden on coal firms above a certain size ranged from 11 percent to 13 percent. For example, the real VAT burden on coal products is 12.92 percent in Anhui, 12.39 percent in Henan, 10.87 percent in Shaanxi, and 10.98 percent in Hebei, according to the China Coal Industry Association. In addition to taxes, there are as many as ninety-two types of fees that coal firms are required to pay, including forty administrative and institutional fees, six operational fees, and forty-six railroad, port, freight, and miscellaneous fees.[17] For example, in Shanxi, VAT accounts for approximately one-third of total levies on coal, other miscellaneous taxes and fees account for nearly two-thirds, and resource taxes make up less than 2 percent.

This additional revenue can then be plowed back into subsidies for clean energy and environmental protection that align with the necessary changes in the economic structure and energy composition.

I. Raise Fees on SO₂ and NOx Emissions One of the policy assumptions in our simulation is that the fees for SO_2 and NOx emissions would be doubled to 2.52 yuan ($0.40)/kg. This is because for a long time, such fees were set too low, which meant that polluting firms were much more inclined to simply pay the fines than to install emissions reduction equipment, which was more expensive. For example, in 2013, the typical SO_2 pollution fee was just 1.26 yuan ($0.20)/kg.[18]

Meanwhile, a survey of multiple thermal power plants found that the unit cost of desulfurization was as high as 3 yuan ($0.50)/kg,[19] and desulfurization costs were even higher for certain companies that did not enjoy electricity price subsidies. Moreover, the NOx pollution fees were generally set under 1 yuan ($0.16)/kg, and were then raised to 1.26 yuan ($0.20)/kg in 2013 in only a few regions, such as Shanghai. This cost differential is clearly a disincentive for power plants to install and run desulfurization and denitration equipment, which would increase the capital input cost by about 5 percent.

Pollution fees that are appropriately set should, in principle, incentivize firms to invest in desulfurization and denitration equipment, which will help to reduce emissions.

II. Increase the Resource Tax and Impose a Carbon Tax The imposition of a carbon tax is another policy assumption in our simulation. The rate at which a carbon tax is levied is typically proportional to the carbon content of fossil fuel products, including coal and downstream petroleum products such as gasoline, aviation fuel, and natural gas. It should be noted that for this model, a carbon tax and a resource tax are treated as the same type of policy because their tax collection methods and economic impacts are basically the same.

However, until recently, China's resource tax was too low and was ineffective in reining in coal consumption. The tax was based on volume, set at 3–5 yuan ($0.50–$0.80)/ton in major coal production hubs, the equivalent of just 0.5–0.8 percent of the price of coal. In December 2014, the government finally approved moving to an ad valorem resource tax for coal, allowing each province to set their rates between 2 percent and 10 percent. As a result, rates announced by Henan, Hunan, Guangxi, Shanxi, Shaanxi, and Inner Mongolia provinces were 2 percent, 2.5 percent, 2.5 percent, 8 percent, 6 percent and 9 percent, respectively, an average of nearly 5 percent.[20] This is the resource tax rate assumed in our model.

Furthermore, the Chinese Academy for Environmental Planning's project team in 2010 recommended levying carbon taxes of 11 yuan ($2)/ton on coal, 17 yuan ($3)/ton on oil, and 12 yuan ($2)/tcm on natural gas, respectively.[21] Based on a coal price of 600 yuan ($100)/ton, an oil price of 4,200 yuan ($700)/ton, and natural gas price of 3 yuan ($0.50)/m³, the proposed rates are equivalent to collecting an ad valorem carbon tax of 1.8 percent for coal and roughly 0.4 percent for oil and natural gas. This is the carbon tax rate assumed in our simulation.

The combined effects of these taxes are an increase of 6.1 percent in the ad valorem rate for the coal industry and a modest increase of 0.4 percent for oil and natural gas. Taxes should be collected at the point of sale—that is, the tax should be collected when the producer sells coal to thermal power plants or to other heavy industries.

III. Boost Subsidies for Clean Energy Finally, we assume that half of the additional fiscal revenue generated from these tax policies would be redeployed to subsidize the expansion of clean energy to raise output, lower relative prices, and spur fuel switching from coal. In the model, this was designed as the equivalent of levying a negative indirect tax on the clean energy sector.

4. MODEL SIMULATION EXPLAINED: AN EXPANDED CGE MODEL
A 135-sector dynamic CGE model was used to perform simulations of a package of policies to curtail coal consumption (see figure 8.4). The three scenarios ("no reform," "partial reform," and "full reform") were separately introduced into the model in the form of external shocks, such as an increase in the resource tax rate, the introduction of the carbon tax, and allocation of subsidies for clean energy.

As figure 8.4 shows, the model encompasses a three-tiered nested production module, which is displayed in two tiers, with the third tier comprising a CES set for domestic and imported products constituting input and a CES set for the different factor types, not shown due to space limitations. The first tier comprises Leontief sets for primary

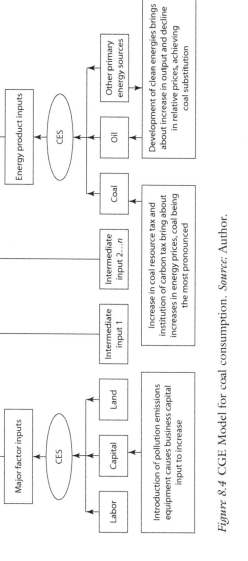

Figure 8.4 CGE Model for coal consumption. *Source:* Author.

factor inputs, N types of intermediate inputs, and energy product inputs. The second tier comprises the three factors of labor, capital, and land, grouped in CES format to form primary factor inputs, with coal, oil, and other primary energy sources grouped in CES format to form the energy product inputs.

Pollution fees are intended to incentivize firms to install emissions equipment by making pollution reduction the optimal option for them. Therefore, in the simulation we approached this problem from the perspective of the capital factor input of firms that use coal, in order to increase the capital inputs and capital depreciation of these firms.

Resource tax and the carbon tax are introduced into the model in the coal-pricing module, causing the price of coal to increase (oil and natural gas prices also rise slightly). The extent of the increase is determined endogenously by the model, based on market supply and demand. Half of the new revenue from the coal resource tax and the carbon tax would be used by the government to subsidize clean energy deployment.

According to our projections, annual coal consumption from 2014 to 2017 will be around 4 billion tons, at a price of 600 yuan ($100)/ton. Assuming an increase in the resource tax rate by 4.3 percentage points to 5 percent, the incremental gain in annual revenue from the resource tax will be about 100 billion yuan ($16.6 billion). We estimate that the annual gain from the carbon tax will be roughly 50 billion yuan ($8 billion). About half of the additional revenues from the resource tax hike and the introduction of the carbon tax (about $12 billion) will be used to subsidize clean energy. This subsidy is enacted in our model by lowering indirect taxes on clean energy sources so that firms are incentivized to replace coal with them.

Impact on the Economy From a macroeconomic perspective, these proposed policies will be mildly negative for economic growth (see table 8.2). Because these policies will catalyze firms to invest in emissions control equipment and clean energy technology, these investments will not directly result in rising output—it is the equivalent of requiring greater capital input per unit of output. Consequently, a slight deceleration in annual output or GDP growth—by about 0.07 percentage points per year relative to the baseline—is an inevitable cost.

However, these policies' impact on the industry structure will be far more significant than on headline GDP growth. The simulation

TABLE 8.2
Cumulative Changes Under the "Full Reform"
Scenario Over a Four-Year Period (%)

CPI	0.01
PPI	0.29
Real GDP	–0.27
Real consumption	–0.06
Real investment	–0.26
Real exports	–0.07
Real imports	0.47

Source: CGE model simulations.

Note: If one assumes that the reform starts in the beginning of 2014 and goes through the end of 2017, then the simulation can be viewed as for the period 2014–2017.

shows that real output and consumption in the coal extraction and processing industry will decline by approximately 9 percent as a result of the policies, meeting the 8 percent target for the 2014–2017 period. Not surprisingly, the industries that will see pronounced negative effects are thermal power, steel, cement, glass, petrochemical, basic chemicals, coking, and rail transportation (see table 8.3). This is because adopting these policies will mean more expensive coal (tax-inclusive) for consumers, higher cost of using coal (due to pollution fee hikes), and less demand for coal transport. On the other hand, subsidies should lead to higher output of clean energy sources, which will in turn bring major changes to the energy composition.

II. Upgrading Industry to Reduce Coal Consumption

1. CURRENT STATE OF CHINA'S INDUSTRY STRUCTURE AND ITS CAUSES Our model indicates that even if the resource tax and pollution fees are raised substantially, a carbon tax is introduced, and subsidies on clean energy are increased, coal consumption will be reduced by only a cumulative 9 percent compared to the results in the "no reform" scenario. Over the longer term, coal consumption must be further slashed. What can be done?

TABLE 8.3

Changes in Real Output of Major Industries (%)

Thermal power generation	-11.77	Gas production and supply	-0.31
Coal extraction and processing	-8.95	Metal product	-0.27
Geological survey	-0.97	Nonferrous metal smelting and alloy manufacturing	-0.25
Oil and nuclear fuel processing	-0.94	Ferrous metal mining and processing	-0.25
Basic chemical raw materials manufacturing	-0.58	Construction	-0.25
Mining, smelting, and specialized construction equipment manufacturing	-0.54	Water production and supply	-0.23
Coking	-0.48	Financial services	-0.20
Road transportation	-0.42	Food services	-0.16
Graphite and other nonmetal mineral product manufacturing	-0.42	Lodging	-0.15
Specialized chemical product manufacturing	-0.42	Agriculture	-0.10
Rail transportation	-0.42	Real estate	-0.09
Synthetic materials manufacturing	-0.41	Public management and social organizations	-0.01
Rail transportation equipment manufacturing	-0.37	Oil and natural gas extraction	-0.01
Steel	-0.36	Clean energy power generation	48.89
Cement, lime, and plaster manufacturing	-0.31		

Source: CGE model simulations.

In our view, long-term coal consumption trends will be determined by the extent to which the entire industrial structure is upgraded. At present, industry as a proportion of China's GDP is notably higher than that in developed economies and even exceeds levels in other BRIC countries at a similar stage of economic development (see figure 8.5). Specifically, heavy industry accounts for more than 70 percent of all industry in China, and as noted previously, coal consumption intensity in heavy industry is nine times that of the services sector (light industry is four times that of services).

Such a disproportionately high level of industry naturally means higher pollution emissions. Therefore, in the process of transforming the economic structure, reducing industry's role in the Chinese economy while simultaneously developing the services sector should be a central component of tackling PM2.5 pollution.

How Did Industry Become So Dominant in the Economy? While industry's dominance in the economy is partly owed to the fact that China is undergoing a rapid phase of industrialization, a series of policy distortions are also responsible for the current structural composition of the Chinese economy. Thus, addressing the industrial structure requires untangling several key distortions.

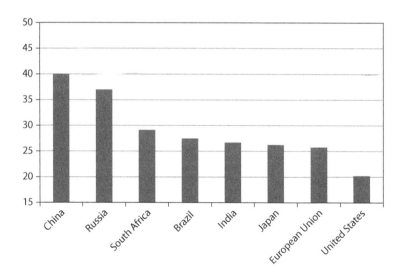

Figure 8.5 Industry as proportion of GDP in various countries, 2011 (%). *Sources:* World Bank; NBS.

First, as discussed, industrial land prices are too low—only about one-eighth the price of commercial and residential land. At the end of 2012, the average price of residential land in 105 major cities in China was 4,620 yuan ($745)/m², while industrial land averaged 670 yuan ($110)/m², just 14 percent of the cost of residential land.[22] This industrial to residential land price ratio is much lower than that of many other countries. In Tokyo, for instance, that ratio is 67 percent, and even in other Japanese cities, such as Fukui, Hyogo, and Fukuoka, where the ratio is lower, industrial land still costs more than one-third of residential land.[23] This price ratio is about 20–30 percent in the United Kingdom, and as high as 60 percent in Taipei, Taiwan.[24]

The low cost of industrial land—in some cases, no cost at all—has led to irrational behavior in terms of land supply, which has essentially subsidized the massive expansion of industry. Currently, in 40 large- and medium-sized cities, land zoned for industrial use accounts for 40–50 percent of the total, but only 20 percent for residential land.

Second, the indirect tax burden on certain services industries is excessive, according to several studies. A quantitative analysis by Ping Xinqiao (2011)[25] indicated that in 2008, in eleven of the more economically developed coastal provinces, the indirect tax burdens (indirect tax/gross value added) on services sectors such as logistics, IT and software services, real estate, leasing and financial services, and R&D, among others, were all higher than the indirect tax burden on industrial firms above a designated size. In addition, Luo Minghua (2010)[26] analyzed data for 2005–2008, concluding that the overall industry tax burden on China's services industries (tax revenue as a proportion of industry value added, including direct and indirect taxes) had surpassed the tax burden on manufacturing industries in 2007. Finally, Tang Dengshan (2012)[27] used input-output table data to extrapolate VAT equivalents for various industries and found that the tax burden on China's construction industry and lodging and food services industry was significantly higher than the average tax burden on all industries.

Many developed countries have adopted uniform tax regimes for all industries, which do not favor services over manufacturing, or vice versa. Some countries, such as Australia and New Zealand, go even further by offering certain preferential taxes for the education, health-care, transportation, financial services, and insurance sectors. China, too, should lower the tax burden on its own services industries.

Although current pilot programs to convert from the business tax to a VAT are intended to address this problem, the effort may result in perverse outcomes. This is because manufacturing industries have already been subject to the VAT in the past, and after converting the business tax to a VAT for services, manufacturing industries can obtain more deductible invoices and actually further reduce its tax burden.

According to the Liaoning provincial branch of the State Administration of Taxation, after the pilot program of tax conversion was completed, "secondary industries stand to become one of the primary beneficiaries."[28] In Jiangsu province, the manufacturing sector experienced the biggest tax cut in a five-month pilot program to convert a sales tax to VAT.[29] Of course, data points are still limited, but these illustrations suggest that there certainly is a risk that, if not properly designed or implemented, tax reforms could end up giving manufacturers an even larger tax cut rather than helping the services industry.

Third, low resource taxes and delays in levying a carbon tax have allowed heavy industry to benefit from low energy prices. Although the Chinese government has been intensely discussing a carbon tax, it has not yet materialized. Meanwhile, the average resource tax on energy products, until recently, was set at only 2 percent,[30] far lower than the rate found in the United States (8–16.7 percent) and Australia (10–12.5 percent).[31] Low resource taxes have historically encouraged rampant development of resource-intensive heavy industry.

2. STRUCTURAL ADJUSTMENT'S EFFECTS ON LONG-TERM COAL CONSUMPTION Now that the policy distortions have been discussed in detail, we turn to simulating the intensity of policy changes required to catalyze substantial structural adjustment and to further reduce coal consumption in the medium to long terms. In addition to the three key policies, certain nonpolicy factors that affect the industrial structure must also be considered.

These nonpolicy factors include changes in the demographic structure and demand elasticity. Both theoretical and empirical evidence demonstrate that an aging population increases a country's consumption rate and lowers its investment rate, which is accompanied by a decline in the proportion of secondary industries.[32] The demand elasticity factor refers to the following: as the penetration of physical goods reaches a certain level, consumer demand elasticity for those goods will gradually decrease, and demand elasticity for services will increase.

We used a 135-sector dynamic CGE model to simulate the effects of the three policy factors on the economic structure. The assumptions for the simulation included:

1. Within ten years, measures would be adopted to raise the price of industrial land by 200 percent. This could be accomplished through decreasing the supply of industrial land or increasing the minimum price for industrial land transfers.
2. The effective indirect tax rate on services industries would decrease by 2 percentage points.
3. Resource tax rates (including those for coal, oil, nonferrous metals, iron ore, nonmetal minerals) would rise an additional 10 percentage points on top of the 5 percent tax rate assumption in the "full reform" scenario. At the same time, the carbon tax would be doubled on top of the assumption in the same scenario—that is, the ad valorem carbon tax for coal would increase from 1.8 percent to 3.6 percent and from 0.4 percent to 0.8 percent for oil and natural gas.

The simulation found that as a result of the policy changes, secondary industry as a proportion of GDP will decline by approximately 1 percentage point and tertiary industry (services) as a proportion of GDP will also increase by 1 percentage point.

We now turn to the question of how the secular trend of an aging Chinese population might affect the economic structure, which can be determined by simulating a comparison of no change in demography versus changes in the demographic structure based on population projections. As such, we found that due to population factors, the proportion of secondary industry will decline by 1.5 percentage points and the proportion of tertiary industries will increase by 1.4 percentage points.

Combining the two sets of results shows that implementing the three reform policies, together with changes in demographics, will lead to notable changes in China's economic structure, with secondary industry as a proportion of GDP declining about 3 percentage points and tertiary industry rising about 3 percentage points.

Many studies show that the economy's demand elasticity would change as incomes rise, because consumers tend to consume more services, such as healthcare, education, travel, and entertainment, when the penetration rate of goods, such as apartments and cars, is sufficiently high.[33] We can derive from these studies that, due to the changes in demand elasticity, secondary industry as a proportion of GDP could fall by an additional 6 percentage points from 2013 to 2030. Combined with the results from the two simulations described above, secondary industry as a proportion of GDP should see a cumulative decrease of about 9 percentage points.[34]

Because coal intensity in secondary industry is far greater than that of the tertiary industry, the results above will imply a substantial reduction of total coal consumption.[35]

CHAPTER NINE

Making Green Finance Work in China

Introduction

"Green finance" generally refers to the financing of investment activities that deliver environmental benefits. It often involves policies and institutional arrangements that incentivize and harness private capital to invest in green projects, in areas such as environmental protection, energy efficiency, clean energy, clean transportation, and sustainable construction. Based on our estimates, from 2014 to 2018, China's green industries will need at least 2 trillion yuan ($330 billion) of investment a year, or about 3 percent of China's GDP. While the Chinese government can contribute 10–15 percent of the capital needed for green investment, the remaining vast majority will have to come from private capital. Attracting private capital into green industries is a major challenge at this point, in part because prices do not fully reflect the positive externality of green projects.

China should establish a comprehensive green finance policy framework to address this challenge by: (1) encouraging private capital into green projects to achieve the nation's emissions reduction objectives; (2) allocating funds in accordance with the principle of "the highest capital utilization efficiency for a given emissions reduction target"; and (3) avoiding systemic financial risk.

This chapter focuses on how to establish a green finance framework in China, with the objective of supporting the central government's goal of meeting its emissions and pollution mitigation targets. The creation of such a financing ecosystem is central to any comprehensive long-term strategy to address China's pollution problem.

The first section presents an economic framework for understanding green finance policy and shaping the policy design. Our research shows that under the current pricing system and firm-level objectives, companies are incentivized to overinvest in polluting industries and underinvest in green projects. However, maximization of social welfare requires firms to increase green investments and reduce polluting investments. Four categories of policies, under which the bulk of measures in a green finance framework should fall, are considered to reshape business decision-making and behavior.

The second section details the types of green financial products that are available or can be created, and discusses relevant international experience that can be adapted for China. The third section examines the role that public finance—such as interest subsidies, price subsidies, guarantees, government procurement, and tax exemptions for green bonds—can play in bolstering green finance. The fourth section discusses how regulations and voluntary principles—such as those requiring financial institutions and publicly traded companies to assume environmental responsibility, incorporation of environmental risk factors into credit ratings, and establishment of a system to evaluate environmental costs—can be used to encourage green investing. Finally, we offer ten recommendations for establishing the green financial system in China.

I. Economic Framework for Green Finance

1. The Producer's Problem: Profit Maximization

The classical assumption in microeconomics is that firms pursue profit maximization. Based on a given output price and input costs, firms arrive at an optimal output quantity by searching for a solution to the profit maximization problem. However, the real problem is that the market prices of these output and input goods sometimes do not fully reflect the externalities generated in the process of producing and consuming these products. Therefore, the output quantities determined by firms based on the profit maximization objective do not necessarily maximize social welfare. For example, because coal prices and production costs do not fully reflect environmental externalities, such as air pollution's effect on health, coal output and consumption exceed the level required for maximization of social welfare. On the other hand, compared to the requirement to maximize social welfare, the output prices of clean energy products are too low, resulting in an insufficient supply of clean energy.

A fundamental flaw of classical microeconomics is that the assumption that all firms maximize profits does not hold. A more realistic assumption to be used for justifying "green finance" is that some firms may pursue both profit and "reputation" (or social responsibility) objectives. The objective function of a firm that incorporates social responsibility can be expressed as:

Objective of socially responsible firm = a × profit + b × social responsibility

where profit refers to sales revenue less product costs and taxes.

Suppose that the firm produces two kinds of products: one is a *clean product* and the other is a *polluting product*. Then the profit of the business is specifically expressed as follows:

Profit = profit of clean product + profit of polluting product = {(1 − tax) × price of clean product × output of clean product − cost of clean products} + {(1 − tax) × price of polluting product × output of polluting product − costs of polluting products}

where output costs include the cost of capital, a part of which is the interest rate, as well as labor costs and depreciation.

Suppose that the production function is characterized by diminishing returns to scale, or rising marginal cost of production. If the objective of the firm is to maximize profit only, then using the condition that marginal revenue equals marginal cost of the two products, the equation has a unique solution to obtaining the optimal outputs of the two products. We call these *profit maximizing output of clean product* and *profit maximizing output of polluting product*.

But because externalities have not been internalized, the following problems arise: The profit maximizing output of the clean product is **less than** the optimal output for maximizing social welfare, while the profit maximizing output of the polluting product is **greater than** the optimal output for maximizing social welfare. (Social welfare is defined here as a firm's profit + utility from personal consumption + externality [e.g., health threats from air pollution imposed on third parties is a negative externality]). A positive correlation exists between these health threats and the production and consumption of the polluting product.

So, how can the externalities be internalized such that the output of the polluting product falls and the output of the clean product rises? There are several policy options to solve the firm's problem.

- *Increase the price of the clean product*—for example, by offering price subsidies for clean energy to increase the return on investment in the clean product and concurrently reducing price subsidies (if any) for the polluting product to lower their return on investment.
- *Lower taxes and funding costs for the clean product*—for example, by providing interest subsidies and tax exemption on interest income from bonds, and concurrently raising taxes and other costs, such as funding costs, on the polluting product to lower the return on its investment.
- *Alter the objective function of the firm*—for example, by increasing the weighting of social responsibility.

While the impact of the first two types of policies on reducing output of the polluting product is rather obvious, the third policy type is not usually found in microeconomics textbooks, but could well be a low-cost and effective policy option. For the purposes of this study, social responsibility is defined as a function that is positively correlated with the firm's output of the clean product and negatively correlated with the firm's output of the polluting product.

The incorporation of social responsibility in the objective function of firms already has been manifest in the public disclosures of major global financial institutions and publicly listed companies. For example, Deutsche Bank's 2012 corporate social responsibility (CSR) report states, "We believe our responsibility goes beyond our core business. Progress and prosperity are driving us when we initiate and support educational, social, and cultural projects."[1] General Electric, too, stated that, "As a 130-year-old technology company, sustainability is embedded in our culture and our business strategy."[2]

If the weight b on social responsibility is greater than zero, to a certain extent, it can change the firm's business behavior in the same way that fiscal and other policy incentives can. Specifically, the social responsibility weight b can potentially achieve the same result as the effect on output generated by a price subsidy for the clean product. We used the two types of firm's problem for the production of a single clean product to generate the following expressions:

Objective of a socially responsible firm = a × profit + b × social responsibility = a × ((unit output price) × output − cost) + b × reputation value × output

Objective of a profit maximizing firm = a × ((unit output price + price subsidy) × output − cost)

If the two equations above are to achieve the same objective, then they can be restated as:

a × ((unit output price) × output − cost) + b × reputation value × output = a × ((unit output price + price subsidy) × output − cost)

Simplifying the equations, the condition for perfect substitution between social responsibility and the subsidy becomes:

b × reputation value = a × price subsidy

That is, under the condition described above, a firm's concern for its reputation can replace the government's price subsidy for the clean product in achieving the same behavior and thus the same optimal output of a clean product.

2. The Consumer Problem

Of course, market prices are determined by *both* firms and consumers in market equilibrium. In other words, consumer preferences affect the product price and therefore also determine the scale of the externality. So the problem of how to change consumer preferences to affect market prices and reduce negative externality also needs to be examined.

Consumers pursue maximization of utility from goods purchased, according to the basic assumption in classical microeconomics. Although marginal utility may diminish, a positive correlation exists between the utility received from each product and the volume consumed. In reality, some consumers care not only about the utility derived from consuming the goods purchased, but also about the externalities that such consumption may generate. We call these "socially responsible" consumers. The utility function of a socially responsible consumer can be written as follows:

Consumer utility function = a × utility from consuming products + b × consumer social responsibility

where the utility gained from the product consumed is the utility function in the traditional sense.

Supposing there are only two types of products, this function can be defined as follows:

Utility brought about by products consumed = U (volume of clean products consumed) + U (volume of polluting products consumed)

Consumer social responsibility = c × volume of clean products consumed − d × volume of polluting products consumed

where c and d are both greater than zero.

For some consumers, a product's price and utility may not be the only factors in the purchasing decision, suggesting that these consumers have begun to incorporate a sense of social responsibility in their consumption behavior. For example, they want to know how a product is produced, where it is produced, including which factory produced it, and whether or not the factory is involved in polluting the environment, employing child labor, or misappropriating intellectual property, among other issues. If the product is associated with these issues, even if it is cheaper, some consumers will decide not to buy it. Social responsibility networks, firms' disclosure of environmental information, and the efforts of nongovernmental organizations all helped to increase the social responsibility of consumers. Under certain circumstances, they may boycott products to pressure firms to stop or correct irresponsible behaviors in the production and sales processes.

To illustrate, Greenpeace has organized consumers in various countries to boycott Nestlé products to protest the 2010 destruction of tropical rainforest and peat beds in Indonesia by Sinar Mas, a supplier of raw materials to Nestlé.[3] Consumers pressured Nestlé to announce that it would stop buying palm oil from Sinar Mas. Many other companies, including KFC, Procter & Gamble, and Coca-Cola, have faced similar consumer boycotts.[4]

Although product boycotts are usually temporary, consumers are nonetheless "voting with their feet" and, in the process, promoting corporate environmental responsibility and gaining "social responsibility utility." Relatedly, some consumers in developed countries have begun to exhibit a sense of pride in driving EVs. This mentality is similar to the pride some business leaders, such as Bill Gates and Warren Buffett, feel about their philanthropic activities.

A 2011 study that Danish wind power company Vestas commissioned, which surveyed 31,000 consumers in twenty-six countries, found that 90 percent of respondents felt that the proportion of renewable energy consumption should be increased globally, and 50 percent of respondents were willing to pay a higher price for products manufactured using clean energy.[5] This and other studies demonstrate that the utility of a consumer product is no longer the only factor in the consumer purchasing decision.

If the weight b in the objective function of a consumer is greater than zero, one can prove that this consumer's optimal consumption of clean products (polluting products) is greater (less) than the optimal consumption of another consumer whose weight b is zero. In other words, the socially responsible consumer's demand for clean products is *greater than* the demand for clean products of consumers who are not socially responsible (conversely, the demand for polluting products of socially responsible consumers is *less than* the demand for polluting products of consumers who are not socially responsible).

If consumers are socially responsible, then because their demand for the clean product is higher, this will cause the price of clean products to increase, and the result is similar to that of a government price subsidy for the clean product. Likewise, if consumers are socially responsible, then because their demand for the polluting product is lower, this will cause the price of the polluting product to fall, an outcome that is similar to a government cut in the price subsidy or an additional tax or fee on the polluting product.

Addressing the demand-side problem of increasing consumers' sense of social responsibility may include educating young children about environmental consciousness, providing the public with transparent information about corporate environmental performance, establishing environmental role models, and leveraging public opinion to censure environmentally unfriendly consumer behavior, among other options.

II. Types of Green Financial Products

1. Green Loans and the Equator Principles

Green lending generally refers to financial institutions adopting practices that target environmentally beneficial projects while restricting lending to those with negative environmental impacts. Green lending can take the form of project financing, construction loans, and equipment leases that serve corporations, as well as home mortgages and car and credit card loans that target retail customers.

In terms of corporate lending, the widely used Equator Principles are a set of voluntary lending standards that stipulate that borrowers must meet certain social and environmental criteria. Banks that adhere to the principles—commonly referred to as "Equator Banks"—will refuse to provide financing for the projects that do not meet the criteria.[6]

The Equator Principles are significant because they are the first attempt to quantify, clarify, and specify previously fuzzy environmental and social criteria in project financing. They have had profound and lasting effects on participating banks' lending practices. UK-based Barclays Bank—one of the first financial institutions to adopt the principles—has established a relatively comprehensive system for evaluating environmental and social risks. Barclays' system includes: (1) requiring bank lending departments and/or internal rating departments to evaluate the environmental impact of projects; (2) assessing projects that have major potential environmental impacts by a specialized environmental and social risk evaluation department; (3) full-process tracking throughout a project's launch, construction, and operational phases. At the same time, in cooperation with the UNEP, Barclays has provided guidance and best practices experience to financial institutions around the world.

Assisted by the Equator Principles and the efforts of several international agencies, a number of green lending projects/plans targeting corporate financing have emerged. For example, in 2009, US-based Wells Fargo Bank established a "National Cleantech Group" to provide exclusive financing and services to the solar, wind, EV, smart grid, and construction industries. The bank extended loans to these industries totaling $2.8 billion and $6.4 billion in 2011 and 2012, respectively, and plans to provide a total of $30 billion in loans to environment-related sectors by 2020.[7] Other international banks that have established similar dedicated green credit departments include BNP Paribas, Standard Chartered, and Rabobank.[8]

Banks are also focusing on commercial construction loans, promoting lending for development of LEED (Leadership in Energy and Environmental Design) buildings in the United States. The New Resource Bank and Wells Fargo, for instance, have offered developers preferential services such as mortgage and insurance premium waivers for green construction.

Retail green credit for individual consumers is also growing in mortgage lending, car loans, and even credit cards. For instance, Citigroup and Fannie Mae have incorporated a household energy conservation indicator in their mortgage loan applicant credit rating systems,[9] while ABN AMRO has implemented a government-led green mortgage program that offers preferential interest rates to borrowers who proactively adopt environmental protection and low-carbon measures, with interest savings of as much as 1 percent.[10] In addition, Citigroup and New Resource Bank offer "one-stop solar energy financing" that provides preferential loans and arranges for solar panel and water heater installation services for households.

Banks in many countries, including the United States and Canada, now offer lower interest rates for the purchase of low-emissions vehicles and fuel-efficient upgrades. And the "goGreen" car loans introduced by BankMecu in Australia are even more specific, not only differentiating loan interest rates strictly according to CO_2 emissions, but also planting trees on behalf of every borrower to offset all GHG emissions of the vehicles purchased.[11]

As early as July 2004, ABN AMRO's Tendris Holdings introduced the world's first credit card linked to an emissions reduction program: the Visa Greencard.[12] The card issuer calculates monthly carbon emissions for every cardholder and invests in corresponding emissions reduction programs. In 2007, Barclays introduced the United Kingdom's first credit card to "halt climate change and promote green consumption," called the "Barclays Breathe Card." Holders of this card are entitled to benefits for purchasing green products, such as natural gas, clean energy power, and bicycles, and Barclays also promises to donate 50 percent of the interest on these cards to emissions reduction programs.[13]

The World Bank and the International Finance Corporation (IFC) have also compiled a number of operating guidelines that target financing in various industries as a reference for banks in the evaluation and management of such risks. As of now, eighty financial institutions in thirty-five countries have adopted the Equator Principles. The project financing granted by Equator Banks represents a market share of over 80 percent of total international project financing.

Few financial institutions in the Asia-Pacific region have adopted the Equator Principles.[14] In China, while the MEP in 2010 compiled and published "International Experience with Promoting Green Credit: The Equator Principles and IFC Performance Standards and Guidelines," these principles have yet to be officially adopted in the commercial banking sector.

One bank, the Fujian-based Industrial Bank Co. Ltd., has been a leader on this front, however. It was the first domestic bank to launch green credit operations and the only Chinese signatory of the Equator Principles. As of mid-2014, the Fujian bank's outstanding green loans, as defined by China Banking Regulatory Commission (CBRC) guidelines, reached 300 billion yuan ($50 billion).[15] As early as 2005, the Industrial Bank had already established a specialized energy efficiency financing team, and in 2009, it created China's first green finance department, the Sustainable Finance Center. This center encompasses five specialized teams, including project financing, carbon financing, market research, technical services, and Equator Principles review.

Moreover, all branches of the Industrial Bank have environmental and social risk management departments and divisions that promote green financial services. In accordance with the requirements of headquarters, the branches have full-time green financial product managers. The Industrial Bank's "green finance, all-out strategy" program offers seven major products, including financing collateralized future incomes from energy management projects, environmental services, carbon assets, emissions rights, and energy conservation and emissions reduction (CHUEE) programs.

Although the Equator Principles have yet to gain significant traction with Chinese banks, the central government has promulgated a number of regulations and policy guidelines to encourage green lending. In July 2007, the MEP, PBOC, and CBRC jointly issued "Comments on Implementing Environmental Protection Policies and Regulations to Guard Against Credit Risk." In addition, CBRC issued "Guidelines on Loans to Support Energy-Saving and Emissions Reduction" and "Green Credit Guidelines" in November 2011 and February 2012, respectively. These green lending policies aimed to shift loans away from "highly polluting, high energy-consuming" firms, but did not provide enough incentives for financing green projects. At the end of 2015, China already had 8.1 trillion yuan ($1.25 trillion) of outstanding green loans, about 10 percent of total outstanding loans.[16]

2. Green Private Equity and Venture Capital Funds

Large-scale direct green investment is still dominated by internationally renowned financial groups, though that is gradually changing. As far back as 1999, the World Resources Institute (WRI) established its "New Ventures" project, with funding support from Citigroup, which focused on investing in small- and medium-sized enterprises (SMEs) in the environmental industries of emerging markets. Target markets for investment included Brazil, China, Colombia, India, Indonesia, and Mexico. From 1999 to 2012, the WRI project helped 367 SMEs that were "generating notable environmental effects" obtain venture capital (VC) totaling $370 million, reduce CO_2 emissions by a cumulative 2.2 million tons, protect 4.5 million hectares of farmland, and conserve and purify a total of 5.7 billion liters of water.[17]

Wells Fargo invests directly in clean energy through its Environmental Finance Group, and already has over 400 projects valued at $3 billion, primarily concentrated in renewable energy—for example, E.ON Climate & Renewables' 203 MW wind project in Magic Valley and phase three of the Sun Edison solar energy project. These projects generated 16 TWh of electricity in 2012, enough to supply 1.2 million American homes with electricity for a year.[18] In addition, the well-known Climate Change Capital Group engages in comprehensive green industry investing and financing operations. Its private equity (PE) division invests only in companies ranging in size from €5 million ($5.5 million) to €20 million ($22 million), and its investments are limited to clean energy, green transportation, energy efficiency, solid waste disposal, and water treatment projects.[19] Dozens of other international financial firms have launched green PE/VC operations, including Environmental Capital Partners, Generation Investment Management, and the Global Environment Fund.

According to survey data from ZERO2IPO (www.pedaily.cn), from 2010 through 2014, PE firms that have disclosed investment information made 7.4 billion yuan ($1.2 billion) in green sectors including new energy, environmental remediation, and clean tech in China (see figure 9.1). Some of these companies have managed successful exits and IPOs domestically and/or abroad.[20] However, PE/VC as a whole encountered several problems in the green industry in China. First, policy support for the sector is relatively weak, and many green projects get low returns. Second, inadequate infrastructure, such as the lack of grid-connected wind power and solar capacity, has been a constraint on domestic demand growth for clean power. Third, the solar equipment sector has been overly reliant on exports to the EU and US markets, which experienced very high volatility in demand. Finally, investors and consumers alike have yet to become fully aware of the benefits of clean technologies and to develop their preference for green products.

3. Green ETFs and Mutual Funds

Many green financial products with relatively sound liquidity have appeared in financial markets, primarily Exchange Traded Funds (ETFs) and mutual funds based on green indices, also including carbon emissions derivatives. These products have attracted a broad group of institutional and individual investors.

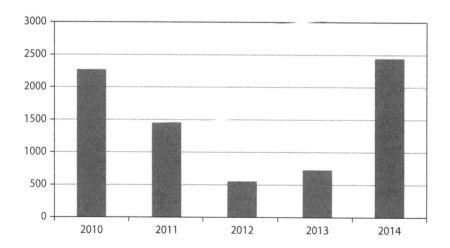

Figure 9.1 Private equity investment in green projects in China, 2010–2014 (million yuan). *Source:* www.pedaily.cn.

Popular green equity indices include the S&P Global Clean Energy Index, which encompasses equities of thirty major global clean energy companies; the NASDAQ Clean Edge US Index, which tracks the performance of over fifty publicly traded clean energy companies in the United States; and the FTSE Japan Green Chip 35 Index, which includes thirty-five listed Japanese companies with environment-related operations. Funds have been created based on each of the indices—for example, the First Trust NASDAQ Clean Edge Green Equity Fund, the iShares Global Clean Energy ETF, and the NIKKO-FTSE Japan Green Chip Index Fund (see box).

China is a latecomer to these efforts, though a few fund products have now appeared on the A-shares market, such as Fullgoal Low Carbon Environmental Protection Equity Fund and the Zhonghai Environmental Protection and New Energy Fund. But they are quite small and their investments are not strictly limited to the environmental realm.

4. Green Bonds

These are low-interest bonds issued for financing green projects. Because of the relatively high credit ratings of the issuers, or because they enjoy government tax exemptions, they are able to fund green projects at relatively low interest rates.

Major financial institutions that have issued green bonds internationally include the World Bank, ADB, European Investment Bank (EIB), IFC, Green Investment Bank (GIB) of the United Kingdom,

---

Examples of Green Indices

Other green indices and related funds include Deutsche Bank's "X-trackers S&P US Carbon Emissions Reduction Fund," JP Morgan Chase's "JP Morgan Environmental Index-Carbon β Fund," Barclays' "Global Carbon Index Fund" that is linked to carbon credit trading on the world's major emissions trading systems, HSBC's "HSBC Global Climate Change Benchmark Index Fund," and the weather contract-based "Global Warming Index (UBS-WGI)," the world's first "Diapason Global Biofuel Index," and "Climate Change Strategy Certificates," all products introduced by UBS.[21]

and the Export-Import Bank of Korea. The investors of these bonds are generally large institutional investors and some high net-worth individuals. Since the first green bond issuance in 2007, the green bond market has expanded rapidly in recent years, with annual issuance reaching $36 billion in 2014.[22]

I. INTEREST RATES The World Bank's first green bond issue, in 2008, was denominated in Swedish kronor and had a par value of 2.325 billion kronor and a maturity of six years. At the time of issuance, the interest rate was 0.25 percent higher than Swedish government treasuries, and investors received an annual coupon of 3.15 percent. The interest rate on the World Bank's green bond in the United States in 2009 was reset every 90 days at the London Interbank Offer Rate (LIBOR) plus 22.5 basis points. A green bond issued by the EIB in 2007 was a zero-coupon bond that paid principal plus a return in the form of a single payment at the end of five years tied to the performance of the FTSE Good Environmental Leaders Europe 40 Index.[23]

II. PURPOSE Proceeds from green bonds fund projects that seek to deliver environmental benefits and address climate change. The World Bank's green bond program supported at least approximately twenty-five such projects, including wind power generation equipment projects in China and Egypt, solid waste management projects in Brazil and Colombia, and reforestation programs in China and Tunisia.

III. MANAGEMENT OF PROCEEDS The proceeds of the World Bank green bonds are put into a dedicated "green account," which is used to manage proceeds from bond issuance. At the end of each quarter, the corresponding funds are remitted into the World Bank's loan pool for distribution by its general loan account to firms.

IV. WHY GREEN BONDS ATTRACT INVESTORS Speaking about green bonds, former California State Treasurer Bill Lockyer stated in 2014, "These investments make sense for the environment and California taxpayers. The return for our pooled money account is outstanding, and we're financing projects that help make a difference in the fight against global warming."[24]

Indeed, several factors make green bonds appealing products. First, green bonds provide a new source of financing for green projects, especially medium and long-term green projects. Second, the green investing themes are attractive to a group of socially responsible investors. Third green bonds have secondary market liquidity, making them easy for investors to sell. Fourth, some green bonds enjoy tax exemptions. Fifth, investing in green bonds issued by institutions such as the World Bank is generally considered lower risk than investing in individual projects or companies. Moreover, the World Bank and other issuers strictly screen the projects in which they invest.

5. *Green Banks*

The United Kingdom established the world's first investment bank that specializes in green investment: the GIB. The creation of the entity was legally mandated as part of the "Enterprise and Regulatory Reform Bill" that the House of Commons passed in May 2012. In the United States, several states have set up green banks, including those established in New York, Connecticut, and Hawaii. (Although they have not formally established independent green lending institutions, some parts of the World Bank, ADB, and Germany's KfW Bankengruppe essentially already operate like green banks.)

The GIB evaluates projects using three standards: first, soundness (sound banking principles); second, leverage effects (the ability to mobilize additional private sector capital); third, green effects (impact). The investment focus is green infrastructure projects of a relatively strong commercial nature. A minimum of 80 percent of its investing activity is targeted toward its preferred sectors, which include offshore wind power, waste recycling, and nonresidential energy efficiency. GIB's investment products include equities, bonds, and guarantees; the bank does not offer loans, VC, or subsidies. It also invites private third parties to jointly invest.

6. *Green Insurance*

Green insurance, which mainly refers to environmental pollution liability insurance, is a risk management tool for project owners to transfer risk to an insurer for potential cleanup costs related to environmental pollution. Environmental liability insurance is generally targeted at the compensation liability that the insured is legally required to assume as a result of polluting water, land, or air. Without green insurance, many firms would not have the financial capacity to provide cleanup funds after accidental pollution incidents. Moreover, green insurance reveals the future environmental costs to project owners and forces them to reassess the viability of projects when the environmental risks are incorporated into overall costs.

Because the European Union has consistently highlighted the "polluter pays" principle in legislation, and in 2004 issued the European Commission's "Environmental Liability Directive" to emphasize pollution liability, green insurance is well developed in European countries. As early as the 1960s, Allianz offered water pollution liability insurance in Germany (German civil law at the time already contained penalties for polluting water). Since then, the scope of environmental insurance has expanded, gradually covering various pollution types including soil contamination. The Association of British Insurers organized insurance companies nationwide to uniformly introduce similar insurance products. When pollution occurs, the insurance covers not only the cost of contamination cleanup, but also covers fines, loss of fixed asset value, attorneys' fees, and medical expenses. In the United States, forty insurance companies now provide green insurance products, and total premiums paid per year are estimated at $2 billion to $4 billion.

A pilot program for environmental liability insurance was launched in China in 2007. Then in January 2013, the MEP and CIRC jointly issued a document directing fifteen pilot provinces to promote mandatory environmental pollution liability insurance in industries involving high environmental risks, such as heavy metals and petrochemicals. It was the first mention of the "mandatory" concept in China, even though the regulatory document remains an "opinion" for now and is not legally binding.

The pilot program has been active for several years, but Chinese firms are still not insuring themselves on a large scale. This is because of many challenges including noncompulsory participation, immature products, unclearly defined operations, high transaction costs, asymmetric information, inadequate pursuit of environmental liability claims, and insufficient disclosure of pollution information.

At this point, legally mandated pollution liability insurance is limited to offshore oil exploration and inland water transport industries in China. For instance, the civil liability insurance for oil pollution is mandated under the "Marine Environmental Insurance Law," the pollution damages liability insurance is stipulated in the "Regulations Concerning Environmental Protection in Offshore Oil Exploration and Development," and the vessel pollution damages policy is mandated under the "Administrative Regulations on Inland Water Transport Safety."[25]

III. Public Finance's "Leverage Effect" on Green Finance

Using public funds to incentivize private investment is an important means of internalizing the externalities of green projects. This approach may include providing interest subsidies on green loans or price subsidies for green output (for example, feed-in tariffs [FiTs] for clean energy electricity), government guarantees for green projects, granting tax exemptions for green bonds, creating state-backed green banks or investment funds, and making public investment to mobilize private capital to invest in green industries.

For instance, in its "Public Finance Mechanisms to Scale Up Private Sector Investment in Climate Solutions," the UNEP states that "public finance mechanisms are a part of a program to solve environmental issues; one dollar spent of public funds can leverage in the range of three to fifteen dollars of private capital."[26] It estimates that $10 billion in public funds can mobilize up to $100 billion in private capital to invest in green industries. The experiences of various countries illustrate how public spending can be used to catalyze private green investment.

1. Providing Interest Subsidies on Green Loans

One of the important features of Germany's green credit policy is participation of the government, which owns a controlling share of Germany's policy bank, KfW Bankengruppe. The bank plays a leading role in financing the growth of SMEs, particularly those in the environmental and clean energy sectors. The bank has established a number of loan products, such as "KfW environmental loan project," the "KfW energy efficiency project," and the "KfW energy capital transfer program," which receive interest subsidies from the federal government.

The first two programs target German and non-German SMEs, as well as individuals, in the environmental and clean energy realm.

Benefits include loans of up to ten years, highly preferential loan rates (with some individual projects receiving rates of less than 1 percent), lump-sum repayments of 100 percent of the loan proceeds, and no repayments for the first three years. The third program is targeted at medium and large enterprises in Germany, and it uses preferential terms to encourage loan applications for installing energy-saving equipment and innovative new energy projects. Loans are extended by KfW in cooperation with other commercial banks.[27]

The German bank has worked with China's CITIC bank to introduce the first stage of green credit projects in 2012. KfW provided project funding with interest subsidies from the government, and China's MoF played the role of the "borrower" to support Chinese SMEs on their energy-saving and emissions reduction projects (MoF lent money and both KfW and CITIC served as financial intermediaries). The KfW mechanism not only allows green industries to obtain financing at preferential terms, it also mobilizes private sector participation through cooperation with other commercial banks.

2. Government Provision of Green Loan Guarantees

A UK study on SME financing[28] argued that the government is not the best candidate to decide which SME can get financing. Therefore, the government must call on the extensive human capital, skills, and experience in the private sector to make lending decisions. The UK government has elected to use the "loan guarantee program" to support SMEs, particularly those in the environmental and clean energy sectors. A company's potential environmental impact should serve as an important criterion in determining the percentage of the final guarantee and repayment. Such policy intervention tools are highly effective and can optimize capital allocation at relatively low cost.

3. Price Subsidies

Currently used in more than fifty countries, a FiT allows governments to offer clean energy firms or individual investors a price at which the state guarantees long-term purchases of their output, so that investors are able to get stable, and likely better, returns. Because return on investment is a major factor affecting market growth, a FiT can quickly and effectively promote the development of clean energy via market mechanisms. The typical terms of a FiT range from ten to twenty-five years.

FiTs are most commonly used in the solar energy industry. For example, in the German Renewable Energy Act promulgated in 2000, the government guaranteed an electricity price of €0.35–€0.50 ($0.38–$0.55)/kWh for all new grid-connected solar power. This meant that the national power grid could enjoy a twenty-year purchase guarantee at a fixed price.[29] Germany's FiT policy allowed solar power companies to spread additional costs evenly to all customers.

Consequently, solar energy as a proportion of total power generated in Germany increased from less than 0.1 percent in 2003 to 5.3 percent in 2012, and the additional cost passed on to end users was only €0.036 ($0.04)/kWh.[30] In addition, to lower cost, the purchase price for newly installed solar capacity has decreased each year. In 2013, for the latest installed solar power generation, the FiT has fallen to €0.10–€0.14 ($0.11–$0.15), with the repurchase term maintained at twenty years (it is a common practice to gradually lower the FiT subsidy to avoid overprotection of the industry). Thus, after the initial stage, the clean energy industry has also begun to gradually face competition from traditional power generation, and the firms involved are maximizing cost-saving efforts while helping the public utility system lower the power tariffs.

FiTs have also been adopted in Spain, Australia, Brazil, Greece, Portugal, South Korea, and Singapore, as well as in a number of states in the United States. The United Kingdom's approach has also harnessed consumer power. That is, British consumers who install solar panels in their homes and use the electricity generated are entitled to subsidies without connecting to the power grid.

4. Tax Exemption for Green Bonds

The laws in most countries require that interest income from bonds be subject to income tax. However, to attract investors in green bonds, in 2013 Massachusetts became the first US state to issue tax-exempt green bonds. Proceeds from the bond issuance are used for building environmental infrastructure. In the corporate bond realm, as early as 2004, the US Congress approved a $2 billion tax-exempt bond program to finance new energy infrastructure.

5. Government Funding of Green Banks

The 2012–2013 annual report of the UK GIB states that direct investment by the bank totaled £635 million ($961 million), with private third-party investment of £1.63 billion ($2.47 billion), which translates into a private capital mobilization ratio of nearly 3:1. That is, for every pound of GIB investment, nearly 3 pounds of private capital was mobilized to support investment priorities. For some individual projects, this ratio was as much as 9:1 (see table 9.1). Because GIB initiated these projects, the private sector viewed the bank's involvement as a partial guarantee, which helped mitigate or spread the project risks. Moreover, GIB involvement lowered transaction costs for private investors because the bank conducted due diligence during the early stages of the project.

IV. Developing Legal and Institutional Capacity for Supporting Green Investment

In addition to providing fiscal incentives, a number of legal and institutional arrangements can mobilize and incentivize private investment in green industries. These arrangements do not necessarily require much public spending, and they can help lenders and investors raise their preferences for green projects and reduce their propensity for investing in polluting projects.

1. Establishing Lenders' Environmental Liability

In 1980, the Comprehensive Environmental Response, Compensation, and Liability Act was introduced in the United States. Pursuant to this act, banks may be held liable for environmental pollution caused by their borrowers and may be required to pay remediation costs. The conditions were as follows: if a lender participated in the business operations, production activities, or waste disposal activities of a borrower that caused pollution, or if it held property rights in facilities that caused pollution, it must assume corresponding liability. This is known as lender liability.

TABLE 9.1
GIB Invested Projects, 2012–2013

Investment project	Industry	Date	GIB investment (£ millions)	Total investment (Private + GIB, £ millions)	Private capital mobilization ratio	Reduction in GHG emissions (hundred tons)
Foresight	Waste/biomass energy	Nov 2012	50	100	1:1	16
Greensphere	Waste/biomass energy	Nov 2012	30	60	1:1	16
SDCL	Energy conservation	Nov 2012	60	100	1:1	n/a
Equitix	Energy conservation	Nov 2012	50	100	1:1	n/a
Drax	Waste/biomass energy	Dec 2012	100	990	1:9	2,500
Walney	Water	Dec 2012	46	224	1:4	51
Wakefield	Waste/biomass energy	Jan 2013	30	122	1:3	34
Gloucester	Waste/biomass energy	Feb 2013	47	185	1:3	7
Green Deal	M&A	Mar 2013	125	169	1:0.4	82
Rhyl Flats	Water	Mar 2013	47	115	1:1	27
Aviva Fund	Energy conservation	Mar 2013	50	100	1:1	8
Total			**635**	**2,265**	**1:3**	**2,741**

Source: GIB Annual Report, 2013.

In *United States v. Mirabile* in 1985, the US District Court in Pennsylvania heard the first case involving the concept of "lender liability." During the bankruptcy reorganization proceedings of paint manufacturer Turco Coatings, Inc., American Bank and Mellon Bank were charged with owning property that polluted the environment and with having authority on the board of directors that affected the company's operations.[35] Although American Bank was ultimately not required to pay penalties because of insufficient evidence, this case led American Bank to become the first bank to consider environmental policy. In 1986, Maryland Bank and Trust was sued in the US District Court of Maryland for declining the Environmental Protection Agency's request to clean up hazardous waste around a property the bank held as collateral against the borrower's debt. The defendant eventually lost and paid the costs for the site cleanup.[36] Over a hundred similar cases have been tried in the United States since the early 1980s.[37]

In China, however, environmental liability of financial institutions has not been legally codified. Existing regulations remain at the conceptual level and act more like moral suasion than legally enforceable provisions.

2. *Incorporating Environmental Factors in Investment Decisions*

The UN Principles for Responsible Investment (PRI) are an international framework organized by major global investors, with the objective of implementing six major investment principles. On April 27, 2006, former UN Secretary General Kofi Annan, accompanied by global institutional investors, unveiled the PRI at the New York Stock Exchange (NYSE). As of mid-2016, about 1,500 global institutional investors, with combined assets under management of $60 trillion, were signatories of the PRI.

The PRI framework emphasizes the need for investors to consider environmental, social, and corporate governance (ESG) elements in their investment decisions. The six major principles are as follows:

1. We will incorporate ESG issues into investment analysis and decision-making processes.
2. We will be active owners and incorporate ESG issues into our ownership policies and practices.

3. We will seek appropriate disclosure on ESG issues by the entities in which we invest.
4. We will promote acceptance and implementation of the Principles within the investment industry.
5. We will work together to enhance our effectiveness in implementing the Principles.
6. We will each report on our activities and progress towards implementing the Principles.[38]

Progress has already been made to convince investors to adopt the principles under this framework. Investment guidance has been provided to help signatory institutions, especially new ones, to consider ESG factors when investing. In addition, a monitoring institution was established to conduct regular reviews. Currently, more than 20 global financial institutions, including Deutsche Bank, Citigroup, Société Générale, and UBS, have incorporated ESG factors into their project and asset allocation analysis models.

For example, in 2012, Deutsche Bank applied the ESG criteria during the decision-making process on assets valued at €3.7 billion ($4.1 billion). Using the ESG framework, Deutsche Bank established a "Get FiT" program to assist African countries with small hydropower projects and a Desertec program to support solar and wind energy in North Africa and the Middle East.

Moreover, investors have been required to publish annual summaries of their PRI implementation status, and the reports and evaluation documents are made public. The signatory institutions also have been required to share their experiences and establish a network of investors through the Clearinghouse forum. Finally, a dedicated research budget has been allocated to collaborate with academia to study the use of ESG criteria by institutional investors, share case studies, and publish findings.

3. Developing Green Credit Ratings

Incorporating environmental considerations into banks' and credit rating agencies' evaluation of corporate and sovereign credit risk is a recent development. A study by Daniel Hann (2011) demonstrated the existence of a positive correlation between environmental risk and a company's financing costs in the bond market. All other conditions being equal, the greater a company's environmental risk, the higher its corresponding financing costs, indicating that the market has already "priced in" this factor when considering investment costs.[39] The study seemed to validate the relevance of green credit ratings of bond issuers to bond investors.

Some banks and rating agencies have begun to consider environmental risks, a practice often referred to as "green credit rating." Barclays, for example, has created a specialized environmental and social risk evaluation system, which includes loan officers, internal credit teams, an environmental and social risk policy team, and a "reputation council."

A typical loan case involves collaboration between only a loan officer and a credit team, but once a company is deemed to pose potential environmental risks—particularly those in sectors such as mining, electric power, water resources, forestry, fisheries, nuclear, solid waste disposal, petroleum and natural gas, metallurgy, or chemicals—the environmental and social risk policy team becomes involved and provides guidance. If there are major risks that could affect the bank's reputation, the reputation council intervenes and has the final say. Every project is required to undergo an "environmental and social impact assessment," and after a loan is approved, the company must also abide by environmental policies during project implementation—conditions that are written into the loan agreement.

Regarding sovereign credit analysis, the UNEP in 2012 published "A New Angle on Sovereign Credit Risk: Environmental Risk Integration in Sovereign Credit Analysis," which evaluated the potential economic effects of environmental change.[40] Analyzing five countries—Brazil, France, India, Japan, and Turkey—the report concluded that deterioration of natural resources can lead to changes in the balance of trade payments and can thus affect a country's sovereign credit risk. The report maintained that environmental factors should be part of the sovereign credit risk evaluation for all countries.

In terms of corporate credit ratings, Standard & Poor's has already stated the need to consider ESG factors, and has focused on factors related to climate change, carbon emissions, and clean energy. It has incorporated evaluations of the relevant risks into its existing "management and governance credit factors" process.[41]

In China, the CBRC's "Green Credit Guide" states that banks should formulate criteria for evaluating their borrowers' environmental and social risks, perform evaluation and classification of their customers' environmental and social risks, and use the relevant results as an important basis for customer ratings, credit access, management, and withdrawal. It also calls for adoption of differentiated risk management measures in the "three reviews" for loans, loan pricing, and the allocation of risk capital. However, because the CBRC guide is not legally binding, implementation has been uneven in the banking system. A major obstacle is the lack of technical expertise in quantifying the environmental risks in their internal credit rating process.

4. Disclosure of Environmental Information

Globally, at least 20 securities exchanges have issued guidelines for listed companies to disclose environmental information. The contents of such disclosures typically include the types of projects the company operates, the environmental impacts or potential impacts of its investments, efforts the company has made to reduce these impacts, and the company's investments in environmental remediation.

The International Organization for Standardization's (ISO) environmental management framework series ISO14000 are currently the most complete and widely accepted standards for disclosing environmental information. The standards cover six areas: environmental management systems, environmental audits, environmental labeling, environmental conduct assessment, life-cycle assessment, and environmental product standards. Methods for reporting environmental costs at the international level primarily include disclosure notes that accompany financial statements (United States), disclosure in the form of an environmental financial statement (Europe and Japan), and disclosure in the form of an environmental costs report.[42]

According to a 2013 Trucost report, for the fiscal year 2011–2012, all of the 443 UK companies in the FTSE All-share Index disclosed their environmental information in various forms, including annual reports and social responsibility reports that quantified the environmental impacts they had generated. This 100 percent disclosure rate was a significant improvement over the 80 percent rate in the previous year (average disclosure rate in the country in 2004 was only around 37 percent).[43]

The reason the United Kingdom has been the leader in the disclosure of environmental costs is that the UK Association of Chartered Certified Accountants implemented a system of commendation for disclosure of environmental cost information starting in 1992. In 1997, the

Advisory Council on Business and the Environment laid out guidelines for environmental reporting standards.[44]

By 1993, the European Union had mandated environmental information disclosure for its enterprises through regulation,[45] while Scandinavian countries established similar requirements before 2000.[46] The Japanese Ministry of the Environment, too, issued a guideline in 2003 for publishing environmental cost accounting and related indicators and the environmental information of publicly traded companies.[47] The Canadian government's requirements are even more stringent, mandating disclosure by all companies (public and nonpublic) and for firms to submit pollution prevention plans, which are also an important basis for loan evaluations. Summaries of the pollution prevention plans must be submitted to the Department of the Environment and are published for public comment.[48]

In other countries, regulations of securities exchanges are used to disclose environmental information in social responsibility reporting. Exchanges such as the Paris Bourse, the Johannesburg Stock Exchange, Bursa Malaysia, and the Australia Securities Exchange all require publicly traded companies to publish social responsibility reports that comprise information on labor, health, environment, and human rights issues.

In China, the Shanghai Securities Exchange in 2008 published guidelines that stated, "The Exchange encourages companies to promptly disclose information on practices and achievements with regard to the assumption of social responsibility, and to release annual social responsibility reports on the Exchange's website at the same time that they publish their annual reports."[49] At a minimum, the reports should include "efforts the company has made to promote environmentally and ecologically sustainable development, for example, in areas such as preventing and reducing environmental pollution, protecting water and energy resources, ensuring that the area where the company is located is livable, and protecting and enhancing biodiversity in the area where the company is located." Moreover, the MEP in 2010 solicited public comment on another guideline for environmental information disclosure, but to date, it has not been issued.

According to the 2012 results of Rankins CSR Ratings, 644 listed companies on the Shanghai and Shenzhen exchanges had disclosed CSR reports, up roughly 11 percent over the previous year (582 companies). Still, that figure constituted just 26 percent of the total number of listed companies in 2012.[50]

5. Establishing Green Institutional Investor Networks

Numerous networks of green institutional investors have reached social responsibility agreements on green investments, promoting the integration of environmental factors into investment decision-making, and encouraging enterprises to assume social responsibility. In addition to the UN's PRI network, which has over 1,000 members managing about $59 trillion in assets,[51] other examples of investor networks include:

- **The Investor Network of Climate Risk (INCR):** Established in 2003, this network includes 100 large, primarily American investors with $11 trillion in assets under management. Investors in the INCR network are aiming to change their approach to investment as part of the solution to climate change. They intend to invest more in environmental protection and sustainable development projects and to promote this concept both in their home countries and globally. This organization presented an investor action plan at the United Nations in 2012, directing investors to cooperate and learn from each other to accomplish this psychological shift in investment practices. The organization has achieved some successes, including convincing the US Securities and Exchange Commission to require companies to disclose climate change information, protecting California's climate change law, promoting the Corporate Social Responsibility Index, and advocating for securities exchanges in various countries to mandate CSR disclosures.[52]
- **The Institutional Investor Group of Climate Change (IIGCC):** Established in 2001, this network currently has eighty members, including major European pension funds and other institutional investors, with €7.5 trillion ($8.2 trillion) in assets under management.[53]
- **The Carbon Disclosure Project (CDP):** This group collects and publishes carbon emissions data and the resulting commercial risks for 2,500 companies in thirty countries and requests that listed companies disclose more information regarding carbon emissions. It currently represents 822 institutional investors with a total of $95 trillion in assets under management.[54] Individual

investors have also shown enormous interest in green investing. According to a UK DWS survey of 300 individual investors in August 2010, 41 percent of the respondents expressed an interest in investing in some kind of green funds in the coming year, indicating that they hoped to diversify investments and assume more social responsibility (see figure 9.2).

6. Establishing a Carbon Trading System

I. CARBON TRADING: DEFINITION AND MECHANISMS The advantage of carbon trading is that, for a given quantity of emissions reduction, efforts can be concentrated in the firms that have the greatest efficiency and the lowest cost, thereby reducing the cost to society as a whole. A study by Montgomery (1972), for example, showed that of the various methods of carbon reduction, emissions trading has the lowest cost.[55] Assuming a perfectly competitive market, it is not necessary for the government to know the cost functions of each individual pollution source; it only needs to determine the total volume of pollution based on environmental capacity. The market, in theory, is able to achieve equilibrium and minimize costs.[56] Stern (2009) pointed out that, when considering dynamic incentives to reduce emissions, an emissions trading system (ETS) is more effective than other approaches.[57]

According to the International Carbon Action Partnership (ICAP) Status Report 2015, seventeen ETSs are operating across four continents, in thirty-five countries, twelve states or provinces, and seven cities. Together, these jurisdictions produce about 40 percent of global GDP. Among the various ETS platforms, the largest is operating in the European Union, which primarily uses the cap-and-trade model.

Under such a model, authorities set quotas for the carbon emissions of participating firms for a specified period, typically one year. These quotas can either be assigned without compensation or be auctioned off at the beginning of the designated period. At the start of each new period, emissions quotas are reallocated. Authorities in charge of setting and allocating the quotas can design the system so that the total amount of quotas decreases each year to achieve emissions reduction goals.

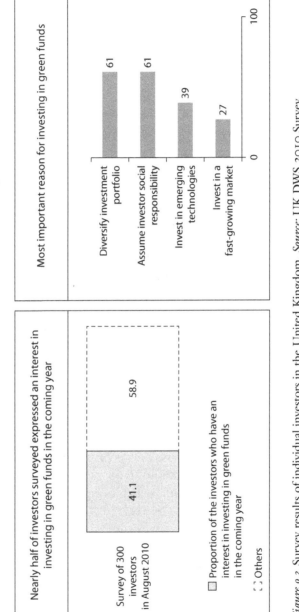

Figure 9.2 Survey results of individual investors in the United Kingdom. *Source:* UK DWS 2010 Survey.

Firms that are more efficient at reducing emissions can sell their unused quotas, and conversely, firms that produce excessive emissions must buy quotas. At the end of the period, firms must verify their quotas and actual emissions volumes with the relevant authorities, and firms that exceed targets can face fines. Because noncorporate institutional investors and individual investors also participate in the ETS, a firm can sell the quota at the time it is received and then buy it back at the end of the period. This means that carbon trading can be a tool for financing.

The EU ETS has more than 11,000 corporate participants from twenty-seven EU member countries, covering 45 percent of total carbon emissions in the entire European Union. From 2005 to 2013, companies covered by the ETS have reduced their emissions by 13 percent. The European Union has also pledged to cut total CO_2 emissions by 1.74 percent each year from 2013 through 2020.[58]

The main problem with the European ETS is that during the 2008 financial crisis, the collapse of demand for carbon quotas led to a fall in carbon prices. The very weak prices failed to generate incentives for emissions reduction during a protracted period of economic recession. This interruption hurt market confidence in the ETS and its function, leading many investment professionals to abandon the market. In response to this failure, the European Commission proposed a Market Stability Reserve (MSR) mechanism to allow the supply of emissions allowances to adjust in response to significant demand fluctuations. That is, allowances can be withdrawn from the market in times of substantial surplus, but kept in the MSR and released in times of extreme scarcity.

II. CARBON TRADING PILOTS IN CHINA In 2011, China's NDRC decided to launch emissions trading pilots in seven provinces and municipalities: Beijing, Shanghai, Guangdong, Shenzhen, Tianjin, Hubei, Chongqing, and Hangzhou. The pilot program period was expected to last from 2013 through 2015. In November 2013, at the Third Plenum of the CCP, the central government unveiled a significant economic reform blueprint that proposed to "create trading systems for energy conservation and carbon emissions rights, pollution rights, and water rights, establish market mechanisms to attract private capital to invest in environmental protection, and promote third-party control of environmental pollution."[59] It reinforced the idea that developing an ETS is an important part of the country's strategy for restructuring the economy.

Similar to the EU cap-and-trade model, Shenzhen was the first pilot to begin actual carbon trading on June 18, 2013. Companies with average annual CO_2 emissions above 5,000 tons from 2009 to 2011 were required to participate in the pilot. The carbon quotas were set at 100 million tons and allocated to 635 industrial firms, including Huawei, Foxconn, and BYD, as well as 197 large state construction groups. The goal was to achieve a 21 percent reduction in carbon emissions in Shenzhen by 2015.[60]

On June 19, 2014, all seven pilot programs became operational, and on average, these pilots covered 40–50 percent of the CO_2 emissions in their respective regions. As of December 1, 2014, the combined market value of Chinese pilot schemes had reached 536 million yuan ($90 million), and the combined trading volume reached 14.4 $MtCO_2$. By the end of 2014, the NDRC announced that, based on the experience of the pilot programs, a national ETS would be launched in 2017.

7. Developing Public Environmental Impact Assessment Systems

All of the previously discussed policies and measures to bolster green investment require quantifying the environmental externalities of specific projects first and foremost. Without the capacity to evaluate the environmental costs or benefits, it is impossible for institutional investors or firms to identify the right green projects in which to invest. For policies such as interest subsidies and tax exemptions that aim to incentivize green investment, the same analytical capacity is necessary within the government to quantify the effectiveness of those policies in driving investment.

Specifically, it is necessary to quantify the emissions generated (or reduced) by each type of production and consumption activity, as well as the externalities that have not been internalized at current market prices. Based on these analyses, the corresponding public spending, financial, and other policies can be designed to internalize these externalities.

A number of practical quantitative methods have emerged to this end. The United Kingdom's Trucost has developed a novel method for quantifying the externalities (see box on next page). Under its proposed "natural capital liabilities" concept, GHG emissions, water resource depletion, and solid waste generation are all considered to erode "natural capital." By establishing an environmental model and integrating calculations performed by a team of experts, Trucost is

Carbon Tax or Carbon Trading?

Academics and policymakers have long debated whether to use carbon taxes (and pollution taxes/fees), an ETS (and pollution rights trading), or a combination thereof to curb carbon (pollution) emissions. In our view, both of these mechanisms have their advantages and can be used in combination if the systems are properly designed. The advantages of a carbon tax lie in its lower administrative costs and relative ease of enforcement. It is necessary to collect this tax from only a minority of upstream industries to cover the consumption of fossil fuels throughout the economy. Collecting a carbon tax can generate additional fiscal revenue, which can then be used to subsidize wind, solar, and other renewable energy, thereby improving the energy composition.

The key advantage of a carbon ETS, on the other hand, lies in the use of market prices to achieve the most efficient allocation of emission rights in order to reduce emissions at the lowest cost. Once a total carbon emissions volume is determined, market participants can reduce emissions through technological innovation or by controlling output, making it easier to accomplish long-term objectives. Once the carbon market is properly established and operational, the information costs to sustain it are relatively low.

At the same time, each of the two methods has limitations. The carbon tax is rigid, hard to adjust, and generates excessive effects on certain industries. A carbon ETS, however, is difficult to design, relatively costly to run in the early stages, and, if not designed properly, has the potential to fail under certain market conditions (recall that carbon prices plummeted during the global financial crisis). Combining both approaches may be necessary in countries with complex economic conditions and inadequate capacity to run each of the two systems optimally.

now able to quantify the corresponding environmental threats or damages, as well as the environmental risks posed by corporate and investor behavior. The results not only include quantitative changes in "natural capital," these changes can also be converted directly into economic values for reference during the investment decision-making process. The company has collected "natural capital liability" from over 4,500 listed companies to form its database. Investors working with the company include Allianz Global Investors, Royal Bank of Scotland, Bank of America, and the NYSE.[61]

V. Establishing a Green Financial System in China

Based on the theoretical framework and international experiences detailed above, and accounting for China's unique challenges, we offer ten recommendations for creating a green financial system in China. The first and third recommendations involve investing public funds

---

Trucost's Environmentally Extended Input-Output (EEIO) Model

Trucost developed the EEIO model to quantify and evaluate environmental costs (externalities). This model covers information on activities in 532 industries. An industry/company's environmental externalities are estimated by evaluating the extent of environmental impacts during various production and trading stages in each industry/firm's supply chain, and these impacts are quantified as monetary values. The advantage of this method lies in the ability to comprehensively tabulate "hidden" environmental costs by tracing back through the supply chain (see fig. 9.3).

Using air pollution as an example, this model determines the impact of pollutants on the ecology and various other linkages and assigns a monetary value to each of the externalities. The specific steps include: (1) using lists of emissions from various countries, the industry emissions volume per unit of output for a particular pollutant is determined (SO_2, NOx, PM, ammonia, carbon monoxide, and VOCs); (2) the output and emissions per unit of output of a particular industry in a specific country are used to estimate the total volume of each pollutant; (3) dose-response functions are used to evaluate the externality (the factors considered include health threats, reduced crop yields, corrosion of raw materials, timber destruction, water acidification, and forest and crop density of the target country); (4) externalities are converted to a monetary value; the primary method used to calculate health threats is the value of a statistical life, reflecting the different levels of productivity and income in each country; (5) all externality values are summed up to arrive at the results.

For example, this model has estimated that the uncompensated environmental costs resulting from air pollution in the Asia region exceeded $200 billion in 2010.

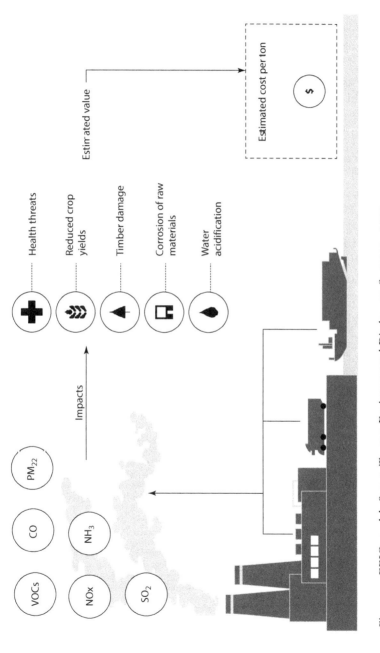

Figure 9.3 EEIO model. *Source:* Trucost Environmental Disclosures Summary, 2013.

to leverage private capital. The reason we have included these "public finance" recommendations is because these measures directly affect the operations of financial institutions and markets. In other words, their function of guiding green investment is realized through the financial system.[62]

The remaining eight recommendations do not require substantial public spending and are considered "nonfiscal" means to incentivize green investing, primarily by influencing financial institutions and market operations. Broadly speaking, all of these actions are aimed at increasing rates of return (or risk-adjusted rates of return) on green projects, so that private investors are more willing to invest in these areas, or at increasing the availability of financing for green projects. Meanwhile, future contingent costs of polluting projects are made more transparent through measures such as mandatory disclosure, green insurance, and environmental education, which also help to limit investments in polluting projects. The effects of these ten policy recommendations are briefly summarized in table 9.2.

1. Establish a Chinese Green Bank

Although many banks are already lending to green projects, the financing provided by the existing financial institutions and the financial markets is insufficient in meeting the expected demand for green investment. By establishing a policy bank—with the government as one of the founders—that exclusively engages in green investment and makes full use of the leveraging effect of green bonds and economies of scale, it will help accelerate the growth in green investment. More specifically:

- Part of the green bank's equity base can come from the state, either via the central government budget or equity injection by PBOC's foreign exchange reserves, with the rest made up of private sources. Private capital could include those from social security funds, insurance companies, other pension funds, and institutions with long-term investment interests. The green bank's capitalization should be sufficient to support hundreds of billions of dollars in green loans and investments each year.

TABLE 9.2
Expected Effects of Various Policies

Policy recommendations	Primary effects and mechanism
1) Develop green banking system	Lowers the cost of funds for green projects and increases availability of funds and efficiency of fund utilization
2) Create green bond market	Provides a new source of funding for green projects
3) Provide interest subsidies on green loans	Lowers the cost of funds for green projects
4) Incorporate environmental risk factors in lending and rating agency decisions	Inhibits polluting investment by making environmental risks more transparent
5) Establish a publicly accessible environmental impact assessment system	Lowers cost of evaluating environmental impact of projects
6) Implement mandatory green insurance in areas of high environmental risks	Determines future environmental risks via insurance premiums, thereby limiting polluting investment
7) Promote development of a carbon ETS	Uses market mechanism to lower the cost of emissions reduction
8) Require mandatory disclosure of environmental information by publicly traded companies and bond issuers	Increases (lowers) green (polluting) investment preferences by increasing public pressure on firms to assume social responsibility
9) Establish a green investor network	Uses pressure from institutional investors to increase the green preferences of invested firms
10) Provide environmental protection education	Increases consumer preferences for green products

Source: Author.

- Some experts have suggested leveraging the green credit operations and professional capabilities already in place at the China Development Bank (CDB) to establish a green investment subsidiary. In our view, both of these options—setting up a new green bank or creating a CDB green investment arm—are open for discussion. However, an independent green policy bank has several advantages: a high-profile green bank can have a very strong effect on the risk aversion of private capital in participating in green investment, specialization can substantially enhance the financial performance and risk management of the bank, and a new bank will allow innovation in capital structure and corporate governance and will avoid many legacy issues with existing institutions.
- Drawing on the practices of the World Bank and the ADB, one of the primary sources of funding for the green bank should come from the issuance of medium- and long-term green bonds. The PBOC's re-lending facility should also be an important financing source for the green bank.
- The green bank should focus on financing green industries and projects, primarily focused on large-scale environmental protection, energy conservation, new energy, and clean transportation sectors. In addition to green lending, the bank could engage in operations such as equity investments and guarantees.
- The green bank should use a variety of means, including attracting private capital as the bank's equity shareholders, bond issuance, guarantees, and co-investment at project levels, to realize the full potential of the government funding's leverage effect. It should strive to leverage 10 yuan ($1.59) of social (private) capital for every 1 yuan ($0.16) of state capital in the green bank's equity base.
- Because of the large scale of the green bank, it will be able to attract the best team of professionals and build capacity for environmental impact assessment, project evaluation, risk management, and financing cost controls. This professional expertise should achieve the greatest economies of scale.
- With respect to loan operations, the green bank should sign on to the Equator Principles and adopt global best practices in disclosing environmental information of its operations and investments.

2. Develop a Green Bond Market

In the past few years, the green bond market has emerged as a new source of financing for green projects in OECD countries and in a number of developing countries. The benefits of the green bond market include: offering a new channel for financing green projects; addressing the problem of maturity mismatch for banks, thereby enabling banks that issue green bonds to increase their green lending capacity; and creating a new asset class that meets the needs of green investors. We recommend that China develop its own domestic green bond market (note that this section was written before the PBOC launched the green bond market in China in December 2015). If well designed, the green bond market has the potential to fund 20–30 percent of needed green investments. The steps toward establishing a domestic green bond market should include:

- Regulators issuing green bond guidelines that specify the basic requirements for issuers, including information disclosure on the use of proceeds.
- Clearly defining the scope of "green bonds" by issuing a Green Bond Catalog.
- Facilitating the development of second-opinion or third-party verification service providers.
- Incubating institutional investors by increasing their preference for green assets, including green bonds.
- Exploring policy options to incentivize the issuance of green bonds, including tax exemption (for interest income on green bonds), interest subsidies, and credit guarantees.

3. Expand and Improve the Interest Subsidy Program on Green Loans

China's existing interest subsidy program for green loans faces a number of issues: (1) it is limited in size; (2) it primarily targets environmental remediation and emissions control projects, such as desulfurization and denitration of coal power plants, which are intended to curb pollution but rarely involve incentivizing investment in other green industries; (3) it targets large projects that are less accessible to SMEs; and (4) it has a complex administration, involving many restrictions on the rates and duration of the subsidies. We believe the interest subsidy program should be improved in the following ways:

- The program on green loans should be significantly expanded, given that they are highly effective in leveraging private capital when compared to direct government investment in green projects.
- Interest subsidies should support not only environmental remediation projects but also additional green projects such as clean energy, energy saving, and green infrastructure. Mechanisms to conduct regular performance reviews and revise the project list should also be established to ensure that the best projects and technologies are eligible for selection.
- The program should not discriminate against SMEs and should be designed to apply different subsidy rates, preferential terms, and approval procedures, depending on firm size.
- Local governments should provide interest subsidies for green projects through local financial institutions.
- Funding for interest subsidies could come from the central or local government budgets and indirectly from revenue generated by pollution fees and resource and carbon taxes.

4. Incorporate Environmental Risk Factors into Project Evaluations

Although China has issued green credit guidelines, their implementation is not mandatory, and the degree to which they have been implemented is uneven across institutions. Therefore, banks' environmental risk controls need to be strengthened in the following respects.

First, the "environmental risk" factor should be formally incorporated into the banks' process of evaluating projects and borrowers. For example, if a project involves a risk of atmospheric pollution, water pollution, or solid waste disposal, a quantitative impact assessment report should be required, as well as solicitation of professional expertise with respect to potential policy changes and the reputational and legal risks the company could face. For certain industries, such as mining, electric power, forestry, fishery, waste disposal, oil and natural gas, metallurgy, and chemicals, these assessments must be made mandatory.

Second, banks should disclose information related to green credit in their annual reports. Third, banks should refer to the Equator Principles to establish dedicated environmental and social risk control departments and carry out full-process management of projects that have an environmental impact.[63] (Full-process management should include credit application review, due diligence, loan approval, and monitoring after the loan has been disbursed.)

In terms of the bond market, third-party rating agencies should provide green credit ratings. That is, credit ratings should include not only traditional risk assessment, but also evaluation of the environmental performance of the projects and bond issuers.

5. *Create an Environmental Impact Assessment System for Public Use*

We recommend establishing a environmental impact assessment system that enables investors to quantify the environmental costs (e.g., emissions)/benefits (e.g., emissions reduction) of projects and that incorporates a complete and continuously updated database of major industries and project types. Such a system should employ calculation methods that are scientific, standardized, comparable, and transparent. The designer of this system should refer to CDB's practice of quantifying the emissions of various projects, as well as to Trucost's "natural capital liability" concept in quantifying the social and economic costs not reflected in market prices.

Because of its enormous social benefits, we recommend that the state or an investors network with government backing undertake the project of creating a public environmental impact assessment system. The results—the technological platform and the associated database— should be considered a quasi-public good that is made available for free or at the lowest possible cost.

This impact assessment system can be used by all investors, including banks, PE funds, brokers, insurance companies, and nonfinancial firms, in their project analysis and investment decision-making. The government can also use this system for policymaking purposes. For example, commercial banks can refer to these environmental cost estimates when making lending decisions, while the government can use this information when determining price and interest subsidies, tax exemptions, and pollution fees.

6. Implement Mandatory Green Insurance in Environmental Risk Areas

Pollution liability insurance—the most important type of green insurance—should be required in sectors where environmental risks are high. We recommend elevating the pilot program on pollution liability insurance that began in 2013 to the national and regulatory level.[64]

In addition to the maritime petroleum exploration and river transportation industries, which are already required to participate in pollution liability insurance, the scope of industries covered should be expanded. While the pilot program opinion suggests inclusion of nonferrous metal mining and processing and smelting, lead-acid battery manufacturing, leather and leather products, chemical raw materials, and chemical product manufacturing, the pilot is largely limited to water pollution and soil contamination risks. Insurance participation should be expanded to include petroleum and natural gas extraction, petrochemicals, thermal power, coal mining and processing, chemical coal processing, steel, cement, plastics, and hazardous chemical product transportation and processing—all sectors that pose significant environmental risks.

Regarding the scope of insurance coverage, regulations can initially target personal injuries and property damages resulting from environmental incidents such as oil spills. Over the long term, the scope of coverage should be expanded to include cumulative damage from environmental pollution—for example, from long-term emissions of exhaust or wastewater.

Property insurance companies should be the primary underwriters of green insurance. These companies may independently establish green insurance units that are similar to catastrophic insurance units.

7. Develop a National Carbon Trading System

Based on the current state of domestic pilot programs and international experience, and in the context of NDRC's plan to launch a national ETS in 2017, we detail several recommendations below.

I. COVER EIGHT INDUSTRIES FIRST The range of industries covered is different in each of the seven pilot programs. We recommend that in the national ETS, seven industries that consume the most energy (more than 100 million tons of standard coal) be incorporated in the initial phase. These industries include coal, steel, chemicals, cement, thermal power, petrochemicals, and nonferrous metals.

In addition, clean energy should be the eighth sector included in the initial phase as well. The reason that clean energy firms should also be included at the outset is because the ETS will become a major source of "subsidy" for them, as clean energy firms will be able to sell quotas in the market. These subsidies from "polluters" to clean energy firms will help accelerate the deployment of clean technology. Firms in the clean energy sector can also pledge their quotas, based on market value under the ETS system, as collateral to secure green loans from banks and issue "carbon bonds" on the bond market.

II. DEVELOP AN "AUTOMATIC STABILIZER" IN QUOTA DESIGN During the early-stage operations of the EU ETS, its rigid quota design, combined with the onslaught of the global financial crisis, led to a period in which excess quotas caused carbon prices to fall to zero. To avoid this problem in China's national ETS system, the possibility of a major economic slowdown should be priced in when setting the quotas. Some kind of "automatic stabilizer" should be built into how the total amount of emissions quotas is determined, so that when the economy dives, the total supply of quotas would also decline. The purpose of this automatic stabilizer is to ensure that prices do not fall too quickly during economic downturns, so that the ETS can consistently provide incentives for emissions reduction even under difficult economic circumstances.

A combination of freely obtained and auctioned quotas can be implemented to better leverage the market and accumulate some revenue. The revenue from quota auctions, together with fines for excessive emissions, can be used to provide subsidies to the clean energy industry and other green projects.

III. ATTRACT PARTICIPATION FROM INSTITUTIONAL AND INDIVIDUAL INVESTORS Lowering barriers to entry for noncorporate investors, including institutional and individual investors, can improve market liquidity and raise public awareness of environmental protection.

Sound liquidity is a necessary condition for carbon prices to effectively reward good performers on emissions reduction.

8. Mandate Environmental Information Disclosure

We recommend that a compulsory disclosure provision be written into China's Securities Law. The CSRC should also issue guidelines that require publicly traded companies and bond issuers to publish environmental information in their CSR reports or as part of their regular financial statements.

The disclosures should include information on emissions and environmental impacts that have been or could potentially be generated by the firms' investments and production activities, as well as the efforts firms have made to reduce such impacts (e.g., by making investments in environmental protection and energy conservation). Disclosure standards should be informed by the widely used environmental cost disclosure information standards in ISO14000. Environmental impacts must be expressed quantitatively and not as abstract qualitative descriptions.

This environmental information may be disclosed as an appendix to the financial statements of a publicly traded company, in the environmental section of a CSR report, or as a separate environmental impact report. All publicly traded companies should meet the disclosure mandate over a period of several years. Over time, environmental disclosures should become one of the requirements for accepting IPO applications. Warnings and penalties should be formulated for public companies that fail to meet the disclosure standard or that falsify information.

In addition, the government should encourage specialized institutions in the private sector to grade and rank the environmental performance, or "green footprint," of publicly traded companies and bond issuers. The grading criteria should include emissions reduction results and compliance with environmental regulations. Consideration can also be given to providing information about the increased or reduced social costs resulting from firms' emissions. This is intended to achieve monitoring effects similar to those of financial audits and can improve the transparency of a firm's environmental performance.

9. Form a Chinese Green Investors Network

Major Chinese institutional investors, such as banks, large insurance companies, investment funds, and brokers, should form their own green investors network. The primary responsibilities of this network should include:

- Promoting the integration of environmental evaluations in the investment decisions of institutional investors, with reference to the ESG factors in the UN PRI, which can be modified to make them more suitable for Chinese investors.
- Supervising companies to assume social responsibility and improve transparency on reporting environmental information.
- Helping the government develop policies to support green investment.
- Sharing best practices on green financial products, services, and information on investment opportunities.

10. Increase Consumer Preferences for Green Products

Consumer social responsibility should be raised through environmental education aimed at increasing their preference for green products and reducing their consumption of polluting products. In other words, a greater number of consumers should be made to lean toward purchasing green products even when prices are slightly higher than prices of nongreen products.

Some approaches to changing consumer preferences include providing environmental education to children and young adults, disclosing environmental information of firms/products to consumers, establishing role models for environmental protection, encouraging NGOs to engage in environmental awareness campaigns and publicity, and using media to censure environmentally unfriendly consumer behavior, among others. Finally, the mass media can publicize green projects and products to help raise awareness and boost market demand.

Notes

Introduction

1. The new standard is for PM10 concentrations to not exceed 70 μg/m³. Refer to report by Zhou Hongchun of the Development Research Center of the State Council, [A high level of importance must be attached to China's PM2.5 pollution prevention and control problems], February 2013.

2. Xie Peng et al., [Study on the exposure-response relationship for atmospheric particulate matter in Chinese populations], *China Environmental Science*, October 2009.

3. Greenpeace and the School of Public Health, Peking University, "Dangerous Breathing—PM2.5: Measuring the Human Health and Economic Impacts on China's Largest Cities," December 2012, http://www .greenpeace.org/.

4. Yuyu Chen, Avraham Ebenstein, Michael Greenstone, and Hongbin Li, "Evidence on the impact of sustained exposure to air pollution on life expectancy from China's Huai River policy," *Proceedings of the National Academy of Sciences* 110, no. 32: 12936–12941, http://www.pnas.org/content/early /2013/07/03/1300018110.

5. Institute for Health Metrics and Evaluation, "Global Burden of Disease Study 2010," *The Lancet* 380, no. 9859, December 2012.

6. Dr. Brian G. Miller, "Report on Estimation of Mortality Impacts of Particulate Air Pollution in London," *Consulting Report P951-001*, June 2010.

7. C. Arden Pope III et al., "Fine-Particulate Air Pollution and Life Expectancy in the United States," *NEJM*, January 2009.

8. World Bank, State Environmental Protection Administration (SEPA), "Cost of Pollution in China: Economic Estimates of Physical Damages," February 2007.

9. Trucost PLC, *Accounting for Asia's Natural Capital*, November 2013.

10. SEPA and National Bureau of Statistics (NBS), *Report on China's Green National Accounting Study*, September 2006.

11. Organisation for Economic Co-operation and Development (OECD), *The Cost of Air Pollution*, May 2014.

12. Chen Tianbing, Wu Jianjun, and Han Jiaye, "An overview of present status of coal burning pollution and treatment technologies," *Coal* 15, no. 2 (2006): 1–4.

13. Chen Danjiang and Yang Guang, "The Comprehensive Development of Coal Chemistry," *China Petroleum and Chemical Industry* 10 (2013): 29–31.

14. Development Research Center of the State Council, Tsinghua University, China Automotive Technology and Research Center, Chinese Research Academy of Environmental Science, "China Motor Vehicle Emissions Control," November 2011.

15. Ruan Xiaodong, "The Challenges of China Petroleum Upgrading," *New Economic Weekly* 4 (2013): 50–53.

16. Zhang Chanjuan and Chen Xiaojun, "Enhancing Motor Vehicle Management and Curbing Air Pollution," *Science & Technology Information* 19, 2013.

17. National Energy Administration (NEA), *Technical Specifications for Managing the Operation of Flue Gas Treatment Facilities at Thermal Power Plants*, 2012.

18. Ministry of Environmental Protection (MEP), *Guide to Best Practical Technologies for Pollution Prevention in the Cement Industry*, 2012.

19. MEP, *Technical Policy for the Prevention of Volatile Organic Compound (VOC) Pollution*, 2013.

20. MEP, *National China V Gasoline Fuel Economy Standard for Vehicles*, 2013.

21. Zhou Hongchun, "A High Level of Importance Must be Attached to the PM2.5 Pollution Prevention and Control Problem," *China Economic Times*, November 2012.

22. Lu Shize, "China Air Pollution: Situation and Solution," MEP public presentation, May 2012.

23. Liu Zixi, "The Impact of Pollution on Chinese Economy," *China High Technology Enterprises* 9 (2013): 122–123.

24. Han Wenke, Zhu Songli, Gao Xiang, and Jiang Kejun, "Large-Scale Pollution: The Urgency of Urban Environment Amelioration," *Price Theory and Practice* 4 (2013): 27–29.

25. Lin Boqiang, "Pollution Can't be Curbed Without Change in Energy Structure," *Comprehensive Transportation* 2 (2013): 42.

26. Jia Kang, "Fiscal Policies and Institutions to Tackle Air Pollution," *Environmental Protection* 20 (2013): 32–34.

27. Chris P. Nielsen and Mun S. Ho, *Clearing the Air* and *Clearer Skies Over China: Reconciling Air Pollution, Climate, and Economic Goals*, MIT Press, 2007 and 2013.

Introduction to Part One

1. Firms above the designated size refer to enterprises with annual turnover of over 20 million yuan ($3 million).

2. According to data from NBS, industrial output from businesses above a designated scale accounts for more than 90 percent of total industrial output.

3. Data and argument are from Chen Xianggui and Jia Wei, "Long-term Projection for China Passenger Car Market," *Automobile Industry Research* 8, 2013.

4. Data taken from announcements and talks by various Ministry of Railway officials and China Railway Corporation managers.

5. Based on the rates used in these government plans and expert forecasts, China's average rail mileage per capita will still be only one-eighth that of OECD countries by 2020, and average subway mileage per capita of large- and medium-sized cities will be only one-fifth that of major international cities.

1. PM2.5 Data, Reduction Model, and Policy Package

Ma Jun and Shi Yu authored this chapter.

1. Data from BP, *Statistical Review of World Energy*, 2012.

2. Data from Beijing Normal University, "Public Transportation and Air Pollution" presentation, 2013.

3. R. Kumar et al., "Air Pollution Concentrations of PM2.5, PM10 and NO₂ at Ambient and Kerbsite and Their Correlation in Metro City," *Environmental Monitoring Assessment*, August 2006; Grazia M. Marcazzan et al., "Characterization of PM10 and PM2.5 Particulate Matter in the Ambient Air of Milan," *Atmospheric Environment*, September 2001. The study by Kumar indicates that the ratio of PM2.5 to PM10 ranges from approximately 0.61 to 0.91; the Marcazzan paper finds that PM2.5 is a major component of PM10, and that the concentration of particulate matter having diameters less than or equal to 2.5 microns is approximately two times that of particulate matter having diameters between 2.5 and 10 microns.

4. http://air.castudio.org. After the writing of this book, the website www.pm2d5.com was converted to air.castudio.org and became a portal that offers air quality information to mobile phone users.

5. MEP, "Zhongguo daqi wuran xingshi yu duice" [Atmospheric pollution in China: conditions and countermeasures], May 2012.

6. National Aeronautics and Space Administration (NASA), "New Map Offers a Global View of Health-Sapping Air Pollution," September 2012.

7. Michael Brauer et al., "Exposure Assessment for Estimation of the Global Burden of Disease Attributable to Outdoor Air Pollution," *Environmental Science & Technology* 46 (2012): 652–660.

8. According to MEP Vice Minister Zhai Qing's speech at the 2015 Annual Conference on Economics of Recycling in Beijing (October 31, 2015), the average PM2.5 level in seventy-four major cities declined by 11.1 percent in 2014, and declined further by 9.4 percent year-on-year in the first nine months of 2015.

9. It is worth noting that the actual decline in the PM2.5 level between 2013 and 2015 occurred faster than our model projected, partly due to: (1) slower than normal wind speed in major cities in 2013; (2) sharper than expected economic deceleration in 2014 and 2015, reflecting weak external demand, lower commodity prices, and property market activities; and (3) stronger than expected government actions to enforce environmental standards and shutting polluting factories and construction sites. Adjusting for the shift in the number of windy days as well as the cyclical factors that influence economic performance, we believe that our projected trend of a gradual reduction in PM2.5 levels over the medium term remains valid.

10. Data from WIND China Macro Database.

11. "Experts pay close attention to the problem of PM2.5 pollution in China", http://www.cq.xinhuanet.com/2013-01/15/c_114375188.htm.

12. Peng Xilong, [Analysis of the characteristics and sources of atmospheric PM10 and PM2.5 pollution in Nanchang]. Nanchang: Nanchang University, 2009.

13. Ye Wenbo, [Analytical study of the sources of PM10 and PM2.5 inhalable particulate matter in Ningbo], 33, no. 9 [Environmental Pollution & Control] (2011): 66–69.

14. Li Qiang, [Analysis of polycyclic aromatic hydrocarbons in atmospheric particulate matter in Jinan, and a breakdown of their sources], Jinan: Shandong University, 2006.

15. Li Jiandong, [Analysis of the chemical composition and sources of inhalable particulate matter in the suburbs of Changsha], Changsha: Central South University, 2009.

16. World Health Organization, "Air quality guidelines for particulate matter, ozone, nitrogen dioxide and sulfur dioxide," 2005, http://apps.who.int/iris/bitstream/10665/69477/1/WHO_SDE_PHE_OEH_06.02_eng.pdf.

17. This assumption of 6.8% annual average GDP growth was used in our model simulation performed in 2013, when GDP growth remained at 7.7%. However, over the past three years (2014–2016), GDP growth decelerated at a pace faster than most economists expected. A more reasonable assumption now (when the English edition of this book is published) appears to be an annual average growth rate of 6.0–6.3% during 2013–2030.

18. In our simulations, we also assumed that the total volume of "other" emissions would remain essentially unchanged over the course of the eighteen-year period. "Other" emissions primarily include sources

such as tobacco smoking, nonbiomass burning, the use of chemical fertilizers and pesticides, cooking, oceans, and forests. Tobacco smoking, cooking, oceans, forests, living habits, and natural conditions are unlikely to undergo material change. The intensity of burning of nonbiomass materials in rural areas and the use of chemical fertilizers and pesticides can be controlled to a certain extent, but total volumes may increase due to economic development. Therefore, real emissions volumes are unlikely to change materially.

2. Environmental Actions: Necessary but Insufficient

Ma Jun and Shi Yu authored this chapter.

1. Many news reports have revealed this phenomenon, most recently in an interview NEA officials in *The Economic Daily* on April 4, 2014, http://paper.ce.cn/jjrb/html/2014-04/04/content_195528.htm.

2. MEP, compilation notes, *Administrative and Technical Specifications for the Operation of Flue Gas Treatment Facilities at Thermal Power Plants*, June 2012.

3. Data from *China Environment Statistics Report 2013*, MEP.

4. If policies such as license plate auction systems and congestion fees are not introduced, we estimate the number of vehicles to rise to 400 million units by 2030.

5. In the last twenty years, fuel efficiency has increased 15–20 percent in European countries.

6. In the last twenty years, the annual mileage driven per vehicle in major European countries has declined 10–15 percent. This is primarily because rapidly developing public transportation systems have led a greater number of people to commute to and from work on subways or trains, which are inexpensive and convenient. Therefore, the decline in the annual mileage driven per individual vehicle has resulted in slower growth in automobile traffic relative to the growth in vehicle ownership. We believe that China should adopt more forceful measures to promote development of urban public transportation systems. This will enable China's automobile utilization rate (measured in annual mileage driven per vehicle) to decline more quickly than that of European countries.

7. Information from European Commission website, Transport and Environment, http://ec.europa.eu.

8. MEP, *2011 Annual Environmental Statistics Report—Exhaust Gas.* http://zls.mep.gov.cn/hjtj/nb/2011nb/201303/t20130327_249976.htm.

9. MEP, compilation notes, *Best Available Technology Guidelines for the Prevention of Cement Industry Pollution*, August 2012.

10. Areas where strict control of atmospheric pollutant emissions is required due to relatively high-density development, a weakening environmental carrying capacity, a relatively small atmospheric environmental capacity, or a fragile ecological environment.

11. MEP, *Technical Policy to Prevent Volatile Organic Compound Pollution*, August 2012.

3. Structural Adjustment: The *What* and the *How*

Ma Jun and Shi Yu authored this chapter.

1. The examples we selected include all major economies for which continuous data are available since 1960 (excluding the United Kingdom). We did not consider extremely small economies (structural changes in these countries and regions are easily affected by distinct factors). We also eliminated countries with transition economies, which in the 1990s experienced substantial economic downturns that also led to dramatic declines in the proportion of heavy industry.

2. See Ma Jun et al., [Quantitative study of China's trade surplus—using a dynamic CGE model to study the effects of aging population, exchange rate and structural reforms on the trade surplus], cited in Ma Jun, [The Traces of Money], China Economic Press, 2012.

3. Tsinghua University Department of Environmental Science and Engineering and SEPA, "Study on Planning for the Control of Motor Vehicle Emissions in China," 2001.

4. Jiang Xuan, "BP: China will become the second largest shale gas producer," *Yicai*, April 29, 2015, http://www.oilzb.com/news/detail/12051.html.

5. Energy Research Institute of the National Development and Reform Commission and the International Energy Agency, *China Wind Power Development Road Map 2050*, 2011.

6. At the time of editing the English version of this book (November 2015), reports said that China's installed solar capacity in mid-2015 had already reached 35.8 GW. This means that the 2020 target of 50 GW is most likely to be exceeded. See "China's solar power capacity to increase by 20 GW per year," China Business Information Net, September 14, 2015, http://www.askci.com/news/chanye/2015/09/14/944396jft.shtml.

7. See note 5 in chapter 2.

8. Zhang Xiliang, dean of the Tsinghua University Graduate School of Energy and the Environment, estimates that the passenger vehicle fleet size will reach 400 million by the year 2030 (http:www.chexun.com/2012-12-13/101632253.html). The China Automotive Energy Research Center at Tsinghua University estimates that the passenger vehicle fleet size will reach 350480 million vehicles by the year 2030, http://qhxb.lib.tsinghua.edu.cn/oa/darticle.aspx?type=view&id=20110626.

9. Calculated based on the research findings of Joyce Dargay et al., "Vehicle Ownership and Income Growth, Worldwide: 1960–2030," *Energy Journal* 28, no. 4, January 2007.

10. A train collision on a high-speed rail line in eastern China in July 2011 killed 39 people and injured 210, see *New York Times*, http://www.nytimes.com/2011/07/27/world/asia/27china.html?_r=0.

11. The meaning of this coefficient is that rail transportation volume can grow 1.2 percent for every 1 percent of growth in railroad length. The reasons for an increase in the rail transportation efficiency coefficient include increases in the proportion of double tracks and departure frequency.

12. "Major cities" refers to cities with population of 5 million or more where subway construction is needed.

13. Due to urbanization, China's urban population growth is higher than overall population growth, therefore the growth rate in urban transportation is slightly higher than the national 5.5 percent transportation growth rate.

4. Enabling Change: Incentives Needed

Ma Jun authored this chapter.

1. Deutsche Bank Corporate Responsibility Report 2012. *Delivering in a Changed Environment.* https://www.db.com/cr/en/docs/CR_Report_2012.pdf.

2. Sustainability at GE, http://www.gesustainability.com/how-ge-works/sustainability-at-ge/.

3. Gallup, *Global Consumer Wind Study 2011*, June 2011.

4. Liu Lin, "The model of industry using land need to adjust," *Securities Market Weekly*, 2013.

5. Data from Japan Ministry of Land, Infrastructure, Transport and Tourism. http://www.land.mlit.go.jp/landPrice/SearchServlet?MOD=2.

6. Data from WIND China Macro Database and Soufun.com.

7. Ping Xinqiao, [The indirect tax burdens of Chinese service industries and sales tax reform], cited in [Forefront of Chinese Financial Policy and Economic Theory], compiled by the Institute of Finance and Trade Economics, Chinese Academy of Social Sciences, 2011.

8. Luo Minghua, [Analysis of the tax burden of China's service sector and countermeasures], [Journal of Guizhou Commercial College], 2010.

9. Zhang Lunjun and Li Shuping, [Study of the industry tax burden of industrial businesses above a designated scale], [Statistical Research], 2012.

10. Tang Dengshan and Zhou Quanlin, [Study of sales tax burdens and value-added tax equivalents], [Henan Social Sciences], 2012.

11. Data provided by Chang Xiujuan, deputy head of Liaoning Taxation Bureau, http://www.chinanews.com/cj/2013/07-29/5097840.shtml.

12. Song Xiaoliang, [Processing and manufacturing sectors benefit from conversion from sales tax to value-added tax], [China Economic News Network], July 29, 2013, http://www.cet.com.cn/ycpd/xbtj/923449.shtml.

13. Jiao Jianling, [Study of long-term and short-term elasticity of coal demand in China], 4 [Journal of Industrial Technological Economics] (2007): 108–110.

14. Liao Yongjin et al., [Study of the cost of flue gas desulfurization installations at thermal power plants], 4 [Electric Power Construction], 2007.

15. Simply put, the marginal utility here is the advantage provided to the car owner by the license plate (e.g., the convenience of doing business as a result of obtaining the license plate allows the car owner to earn a higher income; or the license plate brings the car owner a new sense of happiness/pleasure).

16. Greenpeace, "Analysis of latest 'clean air action plan' introduced by Beijing: Target of 25 percent reduction in PM2.5 concentration in the next five years cannot be realized," September 2013.

17. Ben-Jei Tsuang, Department of Environmental Engineering, National Chung Hsing University, presentation on green building, January 2013.

18. MEP and China Insurance Regulatory Commission (CIRC), *Guidelines on the Pilot Work of Compulsory Pollution Liability Insurance* January 21, 2013, http://www.zhb.gov.cn/gkml/hbb/bwj/201302/t20130221_248320.htm.

5. The Cleanup and Economic Growth

Ma Jun and Xiao Mingzhi authored this chapter.

1. See Ma Jun, [Six reasons and policy implications for the decline in the growth potential of the Chinese economy], [The traces of money], Economic Press China, 2011, http://blog.caijing.com.cn/expert_article-151396-10581.shtml.

2. This projection was made in early 2013 when the Chinese version of this book was written. In 2014, China's coal imports (in volume) declined 11 percent. In the first seven months of 2015, China coal imports (in volume) fell by another 43 percent year-on-year. Data are based on China's General Administration of Customs.

6. Case Study: Shanghai

Li Zhiguo, Ma Jun, Zhang Yan, and Yu Kun authored this chapter. The authors owe a debt of gratitude for the research support provided by graduate students Xiao Xie, Quan Qiwei, Li Xiangyu, and Yang Yukun of the School of Management at Fudan University.

1. Refers to the period from 2013 through 2027, inclusive.
2. Published by Shanghai municipal government on October 18, 2013, http://www.gov.cn/jrzg/2013-10/19/content_2510553.htm.
3. Published by Shanghai municipal government on February 13, 2012, http://cds.nlc.gov.cn/search/htmlflash4Radar?docid=1901702.
4. Transshipment refers to the logistic arrangement for goods to be transported by trucks and/or trains after they arrive via ships at the port. Shore power refers to electricity provided by piers for ships, as opposed to self-supplied power by auxiliary engines and boilers on ships.
5. Fudan University Shanghai PM2.5 Research Team, [Economic policies to reduce PM2.5 emissions in Shanghai], *Boyuan Economic Observations*, December 30, 2013.
6. Considering the differences in the rate of economic structural change, we assumed that PM2.5 emissions reduction in peripheral areas would occur at a rate of 80 percent of the emissions reduction rate in Shanghai.
7. Clean energy is used here in the broad sense, and includes external power generated using clean energy.
8. *BP Statistical Review of World Energy*, 2011.
9. Report dated August 4, 2012, on the website of *Shanghai Business*, http://www.shbiz.cn/Item/182725.aspx.
10. See Singapore Land Transport Authority for explanation of Certificate of Entitlement, http://www.lta.gov.sg/content/ltaweb/en/roads-and-motoring /owning-a-vehicle/vehicle-quota-system/certificate-of-entitlement-coe.html.
11. Mark Goh, "Congestion management and electronic road pricing in Singapore," *Journal of Transport Geography*, 2002, (10): 29–38.
12. Georgina Santos and Jasvinder Bhakar, "The impact of the London congestion charging scheme on the generalized cost of car commuters to the city of London from a value of travel time savings perspective," *Transport Policy*, 2006, 13: 22–33; Robert B. Noland et. al., "The effect of the London congestion charge on road casualties: an intervention analysis," *Transportation*, 2008, 35: 73–91.
13. Todd Litman, "London congestion pricing—implications for other cities," CESifo DICE Report, 2005, 17–21.
14. Shanghai Municipal Statistics Bureau, *Shanghai Statistical Yearbook*, 2011.
15. UK Department of Energy & Climate Change, *Energy Consumption in the UK*, 2015.
16. Fu Qing-yan, Shen Yin, and Zhang Jian, "On the ship pollutant emissions inventory in Shanghai port," *Journal of Safety and Environment*, 2012–5.
17. The Fair Winds Charter proposes requiring oceangoing vessels to switch to low-sulfur diesel while berthing in Pearl River Delta (PRD) waters, and setting up an Emissions Control Area (ECA) in PRD waters. Hong Kong SAR Government, 2011.

18. Hong Kong air quality monitoring network report, HKEPD, 2012.

19. The LegCo Environmental Affairs Panel document, CB(1)625 /11–12(03), 2011.

20. ECAs are classified as either sulfur control areas or nitrogen control areas, and were organized and proposed by the International Maritime Organization and first included in regulations in MARPOL Annex IV.

21. Zhao et al., "Characteristics and ship traffic source identification of air pollutants in China's largest ports," *Atmospheric Environment*, 2013, 64: 277–286.

22. Chen Yimin, Ningbo Municipal Transportation Bureau, 2011, http://www.zgjtb.com/content/2011-06-09/content_189914.htm.

23. Tan Hua et al., "Emission inventory of air pollutants from cargo handling equipment," *Environmental Science and Management*, 2013, 6.

7. Case Study: Beijing

Chen Yuyu, Ma Jun, Yan Se, and Zong Qingqing authored this chapter. The authors would like to thank Peking University graduate students Li Linzhi, Chou Xincheng, and Yan Anran for their research support.

1. In part one we discussed the scenario of reducing the national urban average PM2.5 level from 2013 to 2030, in chapter 6 we discussed the Shanghai case study for 2013 to 2027, and in this chapter we presented the Beijing case study for 2014 to 2028. The shorter time period for the two case studies, as compared to the national level assessment, is due to the greater urgency, and greater resources available to address the air pollution issue in Beijing and Shanghai. The reason that the start of the fifteen-year period of the Beijing case study is one year later than that of the Shanghai case study is because the Chinese version of the Beijing case study was completed in 2014, while the Shanghai case study was completed in 2013.

2. Wen Xu, "Per Capita GDP in Beijing Reaches USD16,278 in 2014," *Xinjin Daily*, January 22, 2015, http://news.sohu.com/20150122/n407994819 .shtml.

3. This model, in its original form, was built by Ma Jun et al. (see chapters 1–3 of this book) to quantify the policy effectiveness of PM2.5 emissions reduction policies on a national scale. Using this model as the basis, we added specific features of Beijing and incorporated many parameters specific to the local economy.

4. Beijing Municipal Government, Beijing Clean Air Action Plan (2013–2017), *Beijing Daily News*, September 13, 2013, http://www.bj.xinhuanet .com/bjyw/2013-09/13/c_117351459.htm.

5. Beijing Municipal Government, Breakdown of Major Tasks in the 2013–2017 Clean Air Action Plan, Beijing-China, September 2, 2013, http://news .xinhuanet.com/local/2013-09/02/c_125301340_3.htm.

6. Beijing Municipal Government,2013–2017 Work Plan for Accelerated Reduction of Coal Burning and Development of Clean Energy, Beijing-China, August 12, 2013, http://zhengwu.beijing.gov.cn/ghxx/qtgh/t1321733 .htm.

7. Beijing Municipal Commission of Urban Planning, Beijing Urban Rail Transportation Construction Plan (2011–2020), Beijing Evening News, October 23, 2012, http://www.tranbbs.com/news/cnnews/news_68545.shtml.

8. Beijing Municipal Environmental Protection Bureau, [Beijing to implement stage five motor vehicle emissions standards effective February 1]. http://www.bjepb.gov.cn/bjepb/324122/451955/index.html.

9. Calvin Quek, "Briefing: Where To Find Data on China's Pollution Problem?" February 25, 2014, http://www.greenpeace.org.uk/newsdesk /energy/data/briefing-where-find-data-chinas-pollution-problem.

10. Detailed Regulations for Implementing the Atmospheric Pollution Prevention and Control Action Plan for the Beijing-Tianjin-Hebei Region and Peripheral Areas http://www.zhb.gov.cn/gkml/hbb/bwj/201309/t20130918 _260414.htm.

11. Zhang Shao, Hebei Provincial Government: Air pollution control action implementation plan in Hebei Province, *Hebei News*, September 6, 2013 http://hebei.hebnews.cn/2013-09/12/content_3477887.htm.

12. Li Tianzhen, "Hebei closed 8347 high-polluting companies, the energy consumption per unit of GDP may decline more than 3%," *Beijing Daily News*, January 13, 2014. http://news.xinhuanet.com/fortune/2014-01 /13/c_125993730.htm.

13. Calvin Quek, "Briefing: Where To Find Data on China's Pollution Problem?" February 25, 2014, http://www.greenpeace.org.uk/newsdesk/ energy/data/briefing-where-find-data-chinas-pollution-problem.

14. On September 27, 2013, Beijing Vice Mayor Chen Gang, spoke before the meeting of the Standing Committee of the Beijing Municipal People's Congress. At this meeting, the Standing Committee considered the "Report Concerning Progress on the Completion of the Proposal to 'Strengthen Urban Planning to Alleviate Traffic Congestion by Addressing the Symptoms as Well as the Causes.'" During his report on the status of traffic congestion control efforts, Chen said, "public transportation accounted for 44 percent of travel in Beijing's central areas in 2012, and rail transportation accounted for 38 percent of public transportation travel." From these figures, it can be extrapolated that public rail transportation accounted for approximately 17 percent of resident travel as a whole in Beijing.

15. "Development Plan for Major Infrastructure Projects in Beijing During the 12th Five Year Plan," Beijing municipal government, November 16, 2011.

16. Liu Xiaoming, "Beijing: the proportion of coal in primary energy consumption structure declines to 20%," from Xinhua Net, March 30, 2015, http://news.xinhuanet.com/local/2015-03/30/c_1114808717.htm.

17. Beijing Municipal Government, 2013–2017 Work Plan for Accelerated Reduction of Coal Burning and Development of Clean Energy. http://zhengwu.beijing.gov.cn/ghxx/qtgh/t1321733.htm.

18. Defined as firms with core business annual turnover of more than 20 million yuan.

19. According to our estimates, under existing policies, such as the "Detailed Regulations for the Implementation of the Atmospheric Pollution Prevention and Control Action Plan in the Beijing-Tianjin-Hebei Region and Peripheral Areas," PM2.5 formed by spillover effects from peripheral areas will decline by approximately 5.5 percent per year through 2017. Because existing policies do not fully consider the problem of inadequate incentives and resources for Beijing's peripheral regions to reduce emissions, without new reform measures, we predict that the annual average reduction in the PM2.5 level will slow to around 3 percent. Our proposed regional compensation mechanism assumes that the mechanism can increase Hebei's effective emissions reduction resources by approximately 50 percent, thus increasing the annual average reduction in the PM2.5 level after 2016 to 4.5 percent. At the same time, because the compensation mechanism requires Beijing to contribute a portion of resources meant to be used locally, we estimate that this will cause the local annual average emissions reduction rate in Beijing to fall slightly. Under this assumption, we estimate that over the course of fifteen years, if the regional compensation mechanism is implemented, the result will be a decline in the PM2.5 level that is 4.1 μg/m³ greater than under existing policies.

20. He Haifeng, "A Survey of Xinanjiang Cross-Provincial Environmental Compensation Mechanism," Sohu blog, August 16, 2012, http://roll.sohu.com/20120806/n349959025.shtml; He Cong, "The Three-Year Pilot Program of First Cross-Regional Environmental Compensation Mechanism," *People's Daily*, December 12, 2014.

21. Hou Jing, "Beijing, Tianjin and Hebei published the details of clean air cleaning plan, within 5 years PM2.5 declines 25%," Enorth net, September 9, 2013, http://news.enorth.com.cn/system/2013/09/19/011321340.shtml.

22. silifayu, "The calculation of the probability to win a license plate in 2014," Beijing Forum, February 25, 2014, http://club.autohome.com.cn/bbs/thread-a-100002-28006543-1.html.

23. Yang, "The car ownership of Shanghai civilian reached 255.19 million, the number of private cars is 183.43 million," Zhongshang net, March 2, 2015, http://www.askci.com/news/chanye/2015/03/02/154919f03a.shtml.

24. Yang, "the car ownership of Beijing civilian reached 532.4 million in 2014," Zhongshang net, February 14, 2014, http://www.askci.com/news/chanye/2015/02/14/93254tl7g.shtml.

25. Beijing small passenger vehicle regulatory information system, http://www.bjhjyd.gov.cn.

26. Shanghai International Commodity Auction Company, Ltd.

27. Tokyo metro subway official website, http://www1.tokyometro.jp/en /corporate/index.html.

28. Zhang Yaowen, "Beijing metro subway plan to construct 12 new lines in 2015," Qianlong net, January 27, 2015, http://beijing.qianlong.com/3825 /2015/01/27/7744@10135226.htm.

29. Firms with core business annual turnover of more than 20 million yuan.

8. How to Deal with Coal

Ma Jun and Xiao Mingzhi authored this chapter.

1. NBS Annual data, Energy, http://data.stats.gov.cn/easyquery.htm?cn =C01.

2. BP, *Statistical Review of World Energy, 2015.* http://www.bp.com /content/dam/bp-country/zh_cn/Publications/2015SR/Statistical%20 Review%20of%20World%20Energy%202015%20CN%20Final%2020150617 .pdf.

3. In this chapter, we define coal gas and coal oil, which produce lower levels of pollution, as noncoal; everything else is considered coal.

4. Rumi Masih and Abul M.M. Masih, "Stock-Watson dynamic OLS (DOLS) and error-correction modeling approaches to estimating long- and short-run elasticities in a demand function: new evidence and methodological implications from an application to the demand for coal in mainland China," *Energy Economics,* 1996, 4.

5. Kong Xianli, [Empirical analysis of the dynamic relationship between coal consumption and influencing factors in China—and a discussion of the asymmetric effects of prices on coal consumption], [Resources Science], 2010, 10.

6. Zhang Lei et al., [Estimates of coal demand elasticity in China: time volatility and regional differences], [Resources Science], 2013, 10.

7. Wei Weixian, [Analysis of China's energy and environmental policy based on a CGE model] [Statistical Research] 2009, 7.

8. Xu Xiaoliang, [The selection of tax rates in resource tax reform: A CGE model's analysis], [Modern Economic Science], 2010, 6.

9. Guo Ju'e et al., [Study on a model to calculate adjustments in the coal resource tax and their effects], [China Population, Resources and the Environment], 2011, 1.

10. Lin Boqiang, Liu Xiying, Zou Chuyuan and Liu Xia, "Resource Tax Reform: A Case Study of Coal from the Perspective of Resource Economics," *Social Sciences in China,* 33, no. 3, August 2012, 116–139.

11. Yin Aizhen et al., [Analysis of mineral resource tax rates based on CGE model simulations], [Finance and Accounting Monthly], 2013, 22.

12. Xu Xiaoliang, [The selection of tax rates in resource tax reform: one CGE model's analysis], [Modern Economic Science], 2010, 6.

13. Guo Ju'e et al., [Study on a model to calculate adjustments in the coal resource tax and their effects], [China Population, Resources and the Environment], 2011, 1.

14. From 2007 to 2012, China's coal consumption grew at an annual average rate of 5 percent versus annual average real GDP growth of 10.1 percent. This implies a coal consumption elasticity of about 0.5. Assuming that energy intensity falls by 20 percent due to industrial structural changes (i.e., the fall in the secondary industry as a percentage of GDP), and energy composition does not change, then our assumption of 5.8 percent annual average real GDP growth would imply coal consumption growth of about 2.3 percent per year from 2013 to 2030.

15. In the PM2.5 control model, if GDP growth and the PM2.5 emissions reduction target are treated as dual macro constraints, based on the potential for emissions reduction from the wider deployment of clean technologies and improvements in the industry and transportation structures, coal consumption can be solved as an endogenous variable.

16. Yu Mengling, "Renewed call for tax cuts on coal producers," *China Energy Daily*, October 29, 2013.

17. Yu Mengling, "Coal firms expect reduction in unreasonable levies," *China Energy Daily*, October 7, 2012.

18. The administration of drainage dues collection, February 28, 2003, http://www.chinalaw.gov.cn/article/fgkd/xfg/gwybmgz/200308/20030800055997.shtml

19. Liao Yongjin et al., "Huodianchang yanqi tuoshu zhuangzhi chengben feiyong de yanjiu" [Study of the cost of flue gas desulfurization installations at thermal power plants], *Dianli Jianshe* [Electric Power Construction], 2007, 4.

20. Xu Ke, "Since December 1st, the coal resource will be ad valorem," China Economy net, October 10, 2014, http://finance.ce.cn/rolling/201410/10/t20141010_3666519.shtml.

21. In the source text, this figure is cited as per cubic meter, but the author believes this was a typographical error. http://cdm.ccchina.gov.cn/Detail.aspx?newsId=1723&TId=1.

22. China's fourth quarter 2012 major urban land price monitoring report.

23. From the Statistics Bureau of Japan.

24. From an article by Liu Lin in [Securities Market Weekly], 2013.

25. Ping Xinqiao, [The tax burden on China's service industries and sales tax reform], [China Social Sciences Press], 2011, 4.

26. Luo Minghua, [Analysis of China's service industry tax burden and countermeasures], [Journal of Guizhou Commercial College], 2010, 4.

27. Tang Dengshan and Zhou Quanlin, [Research on the value-added tax equivalents of the sales tax burden], [Henan Social Sciences], 2012, 6.

28. Enjoy the Dividends from VAT Reform," *People's Daily*, July 24, 2013, accessed at, http://www.mof.gov.cn/zhengwuxinxi/caijingshidian /renminwang/201307/t20130724_968563.html.

29. Song Xiaoliang, [Processing and manufacturing sectors benefit from conversion from sales tax to value-added tax], [China Economic News Network], 2013, http://www.cet.com.cn/ycpd/xbtj/923449.shtml.

30. The average burden for coal is 0.7 percent, and the average burden for oil and natural gas is 5 percent.

31. Li Hengwei and Yang Peigang, [Drawing on international experience in the compensated use of mineral resources and tax reform trends], [Reform], 2013, 7.

32. See Ma Jun et al., [Quantitative study of China's trade surplus—using a dynamic CGE model to study the effects of aging population, exchange rates and structural reforms on the trade surplus], cited in Ma Jun, [The Traces of Money], Economic Press China, 2002.

33. Empirical studies by Buera et al. (2009), Eichengreen et al. (2009), and Development Research Center of the State Council and World Bank (2013) all indicate that tertiary industries as a proportion of GDP in China could increase by 10 percentage points or more in the next dozen years or so. Excluding the impact of policy and demographic factors, the implication of these studies is that service industries as a proportion of GDP could increase by a minimum of 6 percentage points as a result of the demand elasticity factors. See Francisco J. Buera and Joseph P. Kaboski (2009), "The Rise of the Services Economy," NBER Working Paper 14822; Development Research Center of the State Council and the World Bank (2013), "China 2030: Building a Modern, Harmonious, and Creative Society," World Bank publication, March 28; Barry Eichengreen and Poonam Gupta (2009), "The Two Waves of Service-Sector Growth," ICRIER working paper no. 235.

34. By summing up the results of the above-mentioned simulations and the impact of the demand elasticity, the proportion of secondary industries as a percentage of GDP could fall by approximately 9 percentage points during 2013–2030.

35. Note that our estimated reduction in coal consumption is not the reduction in comparison to current consumption volume in absolute terms, but rather the reduction in comparison to the baseline scenario in the future (i.e., the level of coal consumption in ten years, assuming the industry structure and other factors remain unchanged).

9. Making Green Finance Work in China

Ma Jun and Shi Yu authored this chapter.

1. "Corporate Responsibility Report 2012," Deutsche Bank, https://www
.db.com/cr/en/docs/CR_Report_2012.pdf.

2. GE's Introduction, http://www.ge.com/about-us/ecomagination.

3. "Sweet success for Kit Kat campaign: you asked, Nestlé has answered,"
Greenpeace, http://www.greenpeace.org/international/en/news/features
/Sweet-success-for-Kit-Kat-campaign/.

4. List of consumer boycotts, http://www.ethicalconsumer.org/boycotts
/boycottslist.aspx#k.

5. Vestas, Gallup, *Global Consumer Wind Study*, June 2011, 4.

6. According to the International Finance Corporation (IFC), the Equator
Principles (EPs) are a financial industry standard that has been established by
major financial institutions based on the policies and guidelines of the IFC
and the World Bank. They are intended to determine, evaluate, and manage
the social and environmental risks involved in the process of project financing.

7. Wells Fargo Environmental Finance Report, https://www.wellsfargo.
com/downloads/pdf/about/csr/reports/environmental_finance_report.pdf.

8. United Nations Environment Programme (UNEP), Green Financial
Products and Services 2007–2008, 20.

9. UNEP, Green Financial Products and Services 2007–2008, 11.

10. ABN AMRO supports sustainable living, https://www.abnamro
.com/en/sustainable-banking/finance-and-investment/sustainable-living
/index.html.

11. "Be rewarded for being 'green' with a bankmecu goGreen Car Loan",
http://www.finder.com.au/bankmecu-gogreen-car-loan.

12. As of this writing, however, this specific card appears to have been
discontinued.

13. "Barclaycard announces Barclaycard Breathe Easy to help reduce household carbon emissions", http://www.prleap.com/pr/105853/barclaycard
-announces-barclaycard-breathe-easy.

14. "Equator Principles Association Members & Reporting", http://www
.equator-principles.com/index.php/members-reporting/members-and
-reporting.

15. Quarterly reports of China Industrial Bank, http://www.cib.com.cn
/cn/Investor_Relations/.

16. "China Pioneers Sustainable Banking and Shares Experience with
Other Emerging Markets," IFC.

17. "Release: Transition of New Ventures Global Network", http://www
.wri.org/news/2012/09/release-transition-new-ventures-global-network.

18. See Wells Fargo 2015 Corporate Social Responsibility Report, https://
www.wellsfargo.com/about/csr/reports/.

19. Climate Change Capital Private Equity, "Renewable Energy in Italy, investment opportunities and development patterns".

20. ZERO2IPO Research Center, 2007–2012 Annual Research Reports on Chinese Private Equity Investments; 2007–2012 Annual Research Reports on Chinese Venture Capital Investments.

21. UNEP, Green Financial Products and Services 2007–2008.

22. "SEB tops annual Green Bond Underwriters League Table – by a whisker, but Q4 results have Morgan Stanley & JP Morgan on top. Plus we set banks underwriting targets for 2015. Cheeky, yes, but someone has to push them ", http://www.climatebonds.net/2015/01/seb-tops-annual-green -bond-underwriters-league-table-%E2%80%93-whisker-q4-results-have -morgan-o.

23. "EPOS II - The 'Climate Awareness Bond' EIB promotes climate protection via pan-EU public offering," http://www.eib.org/investor_relations /press/2007/2007-042-epos-ii-obligation-sensible-au-climat-la-bei-oeuvre -a-la-protection-du-climat-par-le-biais-de-son-emission-a-l-echelle-de-l-ue .htm?lang=en.

24. "$400m WB Green Bond purchase announced by Standards Board member Bill Lockyer," http://www.climatebonds.net/2014/05/400m-wb -green-bond-purchase-announced-standards-board-member-bill-lockyer.

25. "On the construction of legal system of liability insurance of environmental pollution", *Law Review*, 2015.01 http://www.law.ruc.edu.cn/upic /20150316/20150316090720792.pdf.

26. UNEP, *Public Finance Mechanisms to Scale Up Private Sector Investment in Climate Solutions*, 2009, 4.

27. "Energy efficiency, corporate environmental protection and renewable energies," https://www.kfw.de/inlandsfoerderung/Unternehmen /Energie-Umwelt/index-2.html.

28. Department for Business Innovation & Skills, SME Access to External Finance, 2012, 27.

29. Deutsche Energie-Agentur, www.dena.de/en/.

30. "Germany's Energy Policy: Man-Made Crisis Now Costing Billions," October 30, 2012 http://instituteforenergyresearch.org/analysis/germanys -energy-policy-man-made-crisis-now-costing-billions/.

31. "Buying green! A handbook on green public procurement," 3rd edition, http://ec.europa.eu/environment/gpp/pdf/handbook.pdf.

32. "Federal Acquisition Regulation: Sustainable Acquisition," https:// www.gpo.gov/fdsys/pkg/FR-2011-05-31/pdf/2011-12851.pdf

33. Timothy Simcoe & Michael W. Toffel, "Government Green Procurement Spillovers: Evidence from Municipal Building Policies in California," Working Paper, May 14 2014, http://www.hbs.edu/faculty/Publication%20Files /13-030_a79ab7b7-ad5e-4b80-9e9a-f700baaa9e68.pdf

34. "Notice on the continued development of new energy vehicles promotion and application," http://www.gov.cn/zwgk/2013-09/17/content_2490108.htm.

35. *United States v. Mirabile*, 84-2280 (E.D. Pa. September 4, 1985).

36. *United States v. Maryland Bank & Trust*, 632 F. Supp. 573 (D. Md. 1986).

37. A.C. Geisinger, "From the *Ashes of Kelley v. EPA*: Framing the Next Step of the CERCLA Lender Liability Debate," 4 Duke Environmental Law & Policy Forum, 1994.

38. "Opening NY Stock Exchange, Annan sounds bell for responsible investment," 27 April 2006, http://www.un.org/chinese/News/story.asp?newsID=5542.

39. Daniel Hann, "ESG in the Credit Market," 2012, http://www.idei.fr/fdir/wp-content/uploads/2011/02/hann.pdf.

40. "A New Angle on Sovereign Credit Risk, E-RISC: Environmental Risk Integration in Sovereign Credit Analysis," UNEP, 2012.

41. See Standard and Poor's credit ratings manual, http://www.standardandpoors.com/aboutcreditratings/RatingsManual_PrintGuide.html.

42. Liu Limin and Liu Jingzhi, [International comparison of environmental cost disclosure and lessons to be learned], [Commercial Research], 2008, 12.

43. Trucost, Environmental Disclosures Summary, August 2013.

44. ACBE, "Environmental reporting and the financial sector: An approach to good practice," February 1997.

45. EU, *Community eco-management and audit scheme* (EMAS), 1993.

46. The Swedish Environmental Code, 1998; Norway's Accounting Act, 1998; Danish Environmental Protection Act amended July 1995.

47. Japan Ministry of Environment, Environmental Reporting Guidelines, 2003.

48. Government of Canada, *Pollution Prevention: A Federal Strategy for Action*, 1995.

49. Shanghai Stock Exchange, "Notice on Issuing the guidelines of environmental information disclosure of Listed Companies in Shanghai Stock Exchange," August 15, 2008, http://law.lawtime.cn/d668718673812.html.

50. Lü Qian, "Only 26% of listed companies issued CSR report and 20 companies interrupted issuance," People Net, June 4, 2013, http://finance.people.com.cn/stock/n/2013/0604/c67815-21729710.html.

51. UN PRI fact sheet: http://www.unpri.org/news/pri-fact-sheet/.

52. Investor Network, http://www.ceres.org/investor-network.

53. The investor voice on climate solutions, http://www.iigcc.org/.

54. About CDP: Catalyzing business and government action, https://www.cdp.net/en-US/Pages/About-Us.aspx.

55. David W. Montgomery, "Markets in Licenses and Efficient Pollution Control Programs," *Journal of Economic Theory* 5, no. 3 (1972): 395–418.

56. Ibid.

57. Nicholas Stern, *The Economics of Climate Change: The Stern Review*, Cambridge, UK: Cambridge University Press, 2007.

58. The EU Emissions Trading System (EU ETS).

59. Decision of the CCP Central Committee on "Some Major Issues Concerning Comprehensively Deepening Reforms," November 2013, http://www.china.org.cn/china/third_plenary_session/2013-11/16/content_30620736.htm.

60. China Emissions Exchange, https://ieta.memberclicks.net/assets/BPMR/Shenzhen_Guangzhou/china%20emissions%20exchange_ieta%20bpmr.pdf.

61. Trucost, Environmental Disclosures Summary, August 2013.

62. Other public finance policies that affect green investment (such as the government's tax policies with respect to polluting industries, direct subsidies for clean energy and alternative energy vehicle firms, etc.) are not included in this discussion because they do not necessarily achieve their incentivizing effects through financial institutions. For a detailed discussion of this type of fiscal policy, see the chapter on controlling the consumption of coal.

63. MEP, [Report on the development of green credit in China], Industrial Bank, 2012.

64. MEP and CIRC, *Guidelines on the Pilot Work of Compulsory Pollution Liability Insurance* January 21, 2013, http://www.zhb.gov.cn/gkml/hbb/bwj/201302/t20130221_248320.htm.

Index

ABN AMRO, 228
absolute level, of coal consumption, 199
"Action Plan for Air Pollution Prevention and Control" (State Council), 3
ADB, 232, 256
adjusted human capital (AHC), 5
administrative policy, 101
"Administrative Regulations on Inland Water Transport Safety," 236
Africa, 242
AHC. *See* adjusted human capital
Air Quality Guidelines, of WHO, 2, 30
Allianz Global Investors, 235, 251
American Bank, 241
Anhui, 182–84
Annan, Kofi, 241
Annual Report on the Prevention of Motor Vehicle Pollution (MEP), 45
AP1000 reactor technology, 67
Arrow, 118
Association of British Insurers, 235
Association of Chartered Certified Accountants, 244
atmospheric smog, 133

auction systems, license plate, 43, 70, 88–91, 93, 145–48, 152, 189–91
Australia, 103, 118, 217, 228; FiTs in, 238; PM2.5 in, 27; preferential taxes in, 84, 216
Australia Securities Exchange, 245
automatic stabilizers, 261
automobiles, 116–17; in Beijing, 175–76; future growth in, 34; license plate auction systems, 43, 70, 88–89, 93, 145–48, 152, 189–91; ownership of, 89; private, 22; structural adjustments and, 69–74; reducing emissions of, 43–47; sales of, 104; vehicle fleet size, 141, 144–45, 175, 188–94, 270n6
average energy elasticity, 32–33

bag-filtering dust precipitators, 43
BankMecu, 228
Bank of America, 251
banks, 228; Equator Banks, 229; green, 95–96, 234, 239. *See also specific banks*
Baosteel, 144
Barclays Bank, 227, 228, 232, 243
Barclays Breathe Card, 228
baseline scenario, 32

Beijing, 1, 3, 14, 22, 47, 66,
90–91, 94–95; challenges in
meeting target for: automobiles
and, 175–76; clean energy and,
178; heavy industry and, 179;
peripheral areas and, 172–75;
rail transportation and, 176;
clean energy in, 178, 194–96;
environmental actions in, 167–
72; effects of, 168–70; overview
of, 167–68; heavy industry in,
179, 196–97; introduction to,
163–65; PM2.5 control model,
adapting for, 165–67; regional
compensation and structural
adjustments, 180–98
"Beijing Clean Air Action Plan
(2013–2017)," 167
Beijing Environmental Protection
Bureau, 169
Beijing Infrastructure Investment
Co., Ltd. (BII), 192–93
Beijing Municipal Commission of
Rural Affairs, 169
Beijing Municipal Commission of
Transport, 189
Beijing Municipal Construction
Committee, 83
"Beijing Municipality Provisional
Regulations to Adjust and
Control the Number of Small
Passenger Vehicles," 189
Beijing Municipal People's
Congress, Standing Committee
of, 275n14
Beijing Municipal Statistics Bureau,
163
Beijing Statistical Yearbook, 179
Beijing-Tianjin-Hebei region (Jing-
Jin-Ji), 3, 22, 163, 172, 174,
184–88
"Beijing Urban Rail Transportation
Construction Plan (2011–2020),"
169, 176
Beijing V auto emissions standard,
169

benchmark water quality index
(BWQI), 183
"beneficiary pays" principle, 93,
175
Ben-Jei Tsuang, 93
BII. *See* Beijing Infrastructure
Investment Co., Ltd.
BNP Paribas, 228
bonds: green, 232–34, 238;
municipal, 93, 152
boycotts, of products, 225–26
Brauer, Michael, 21
Brazil, 230, 238, 243
"Breakdown of Major Tasks in the
2013–2017 Clean Air Action
Plan," 168, 175
Bretton Woods, 118
BRIC countries, 215
Buera, Francisco J., 279n33
Buffett, Warren, 226
bureaucratic capture, 20
Bursa Malaysia, 245
BWQI. *See* benchmark water quality
index

Canada, 245
capital: AHC, 5; market-clearing,
125; mobility, 125; natural
capital liability, 5, 250, 251, 259;
price equation, 123; private, 92,
95, 152; venture capital funds,
230–31
carbon: hydrocarbons, 45; taxes,
86–87, 104, 209–10, 212, 251;
trading system, 247–50, 251,
260–61
Carbon Disclosure Project (CDP),
246–47
cargo: dry bulk, 142; transport, 143;
transshipment, 156–58
car lottery system, 189–91
CBM. *See* coal-bed methane
CBRC. *See* China Banking
Regulatory Commission
CCP. *See* Chinese Communist Party
CDB. *See* China Development Bank

CDP. *See* Carbon Disclosure Project
cement, 24, 43
certificate systems, 89
CGE. *See* computable general
  equilibrium
Chen Danjiang, 7
Chen Gang, 275n14
Chen Tianbing, 6
Chen Xiaojun, 7
Chen Yuyu, 4
China Banking Regulatory
  Commission (CBRC), 229,
  243–44
China Coal Industry Association,
  208
China Development Bank (CDB),
  256
China Electricity Council, 41
China Insurance Regulatory
  Commission (CIRC), 98
China Meteorological
  Administration, 68
"China Motor Vehicle Emissions
  Control," 7
China National Coal Group
  Corporation, 208
China National Petroleum
  Corporation, 67
China Railway Corporation, 91
China Shipping Database, 142
Chinese Academy for Environmental
  Planning, 87, 210
Chinese Communist Party (CCP), 78
Chinese People's Political
  Consultative Conference, 30
CIRC. *See* China Insurance
  Regulatory Commission
CITIC bank, 237
Citigroup, 228, 242
civil liability insurance, 235
classic economic theory, 181
Clean Air Action Plan, 168
clean coal technologies, 16, 17, 60,
  116, 205
clean energy, 20, 36, 64, 65f, 115,
  139; in Beijing, 178, 194–96;

positive externalities and, 79; in
  Shanghai, 142, 153–56; subsidies
  on, 87–88, 110, 195, 210–13. *See
  also specific types*
clean production, 33, 35
clean products, 222, 223
Clearinghouse forum, 242
*Clearing the Air and Clearer Skies
  Over China: Reconciling Air
  Pollution, Climate, and Economic
  Goals* (Nielsen and Mun), 9
Climate Change Capital Group, 230
Climate Change Strategy
  Certificates, 232
climate conditions, 32
CNPC, 111, 114
coal, 24, 114–15; Beijing and, 178;
  carbon tax and, 209–10; clean
  technologies, 16, 17, 60, 116,
  205; conversion technology, 7;
  demand structure of, 206–7;
  energy structure and, 59–69; fees
  and, 208, 209; fiscal policies and,
  205–7; gasification technology,
  64; industrial consumption of,
  141, 206–7; industry and, 102,
  213–19; introduction to, 199;
  literature review on consumption
  of, 200–219; PM2.5 trends and,
  204–5; policy recommendations
  and, 207–10; rapid growth in
  consumption of, 14; reducing
  consumption of, 40–42; resource
  taxes and, 85–86, 103, 104, 209–
  10; subsidies on clean energy
  and, 210–13; taxes and, 205–7,
  208; thermal power and, 43
coal-bed methane (CBM), 64,
  66–67
coal-burning boilers, 174, 207
"Coal Reduction and Replacement
  and Clean Air," 170
Coca-Cola, 225
coefficients: dispersion, 30; Shanghai
  and, 135
commodity composites, 128

commodity production price
   equation, 122–23
composite commodity price
   equation, 123
Comprehensive Environmental
   Response, Compensation, and
   Liability Act, 239
computable general equilibrium
   (CGE), 58, 86, 102; basic
   equation set for static, 118–25;
   domestic demand module,
   121; import-export trade
   modules, 122; margin demand
   module, 122; market-clearing
   module, 124–25; price module,
   122–23; production module,
   119–21, 120f; coal and, 200,
   201; economic growth and,
   106–10; expanded, 210–13;
   overview of, 102–3, 117–18;
   policy assumptions in, 103–5;
   simulation results, 106; treatment
   of energy substitute and, 125–28
congestion fees, 43, 70, 91, 93,
   148–49, 191
Connecticut, 234
construction: dust, 51; emissions,
   47–51; growth, 33; projects,
   174
consumer demand, 1
consumer preferences, 224; for
   green products, 263
consumer price index (CPI), 102,
   106
consumer problem, 224–26
consumer products: boycotts of,
   225–26; clean, 222, 223; green,
   16–17, 100, 263; for green
   finance, 227–36; petroleum, 7,
   33, 44, 135; polluting, 222
consumer purchasing power, 149
consumption elasticity, 53
container shipping, 142
corporate lending, 227
corporate social responsibility
   (CSR), 80–81, 223

Corporate Social Responsibility
   Index, 246
cost-benefit ratio, 187
"Cost of Air Pollution, The"
   (OECD), 6
"Cost of Pollution in China:
   Economic Estimates of Physical
   Damages" (World Bank), 5
CPI. *See* consumer price index
credit cards, green, 228
credit ratings, green, 242–44
CSR. *See* corporate social
   responsibility

DBCGE, 118
demand: consumer, 1; domestic
   module, 121; elasticity, 58, 217;
   inventory, 121; margin module,
   122; structure, of coal, 206–7
demographics, 58, 217
denitration, 3, 7, 16, 19, 42, 60,
   166, 205; actual rate of, 43; flue
   gas, 43; in Shanghai, 135
Denmark, 226
Department of Energy, US, 66
Desertec program, 242
desulfurization, 3, 7, 16, 19, 60, 86,
   111, 166, 205; in Hebei, 186–87;
   in Shanghai, 135; of thermal
   power, 40–42
"Detailed Regulations for
   Implementing the Atmospheric
   Pollution Prevention and
   Control Action Plan for the
   Beijing-Tianjin-Hebei Region
   and Peripheral Areas," 172, 184,
   276n19
Deutsche Bank, 60, 81, 223, 232,
   242
Diapason Global Biofuel Index, 232
diesel fuel, 44, 47
dispersion coefficient, 30
domestic commodity, 119, 121
domestic demand module, 121
Dragonomics, 174
dry bulk cargo, 142

dust, 27, 47, 170; bag-filtering precipitators, 43; construction, 51; control of, 33, 35, 168

ECAs. *See* emissions control areas
Ecomagination, 81
economic growth, 212–13; CGE model and, 106–10
economic losses, 4–6
economic structure, 13
economic theory, 77
EEIO. *See* Environmentally Extended Input-Output
EIB. *See* European Investment Bank
Eichengreen, Barry, 279n33
electricity, 111; switching from gasoline to, 160
electric vehicles (EVs), 45, 117, 196, 226
emissions control areas (ECAs), 159, 274n20
emissions per unit, of coal consumption, 199
emissions standards upgrade, 45
emissions trading system (ETS), 247, 249, 251, 261
end-of-pipe controls, 3, 6, 9–10, 30, 35, 52, 75, 103; baseline scenario and, 32; Shanghai and, 136. *See also* denitration; desulfurization
energy: commodity composites, 128; consumption of, 33; elasticity, 32–33, 35, 54–55, 56; future growth in, 34; intensity, 14; renewable, 154–55, 195; structural adjustments and, 55; substitute, treatment of, 125–28. *See also* clean energy
Energy Research Institute (ERI), 68
energy structure, 59–69; hydropower and, 68; natural gas and, 65–67; nuclear power and, 67–68; solar power and, 69; wind power and, 68
enforcement: intensity, 171; ordinary, 168, 172

engine upgrades, 158–60
Enterprise and Regulatory Reform Bill, 234
entitlement systems, 89, 146
Environmental, Social, and Governance (ESG), 99, 241–42
environmental actions, 29, 30; assessing overall results from, 51–52; auto emissions and, 43–47; in Beijing, 167–72; coal consumption and, 40–42; construction emissions and, 47–51; industry emissions and, 47–51; in Shanghai, 136–38
Environmental Capital Partners, 230
environmental impact assessment system, 259–60
environmental information disclosures, 99, 262
"Environmental Liability Directive," 235
environmental liability insurance, 234
Environmentally Extended Input-Output (EEIO), 252
Environmental Protection Agency (EPA), 241
environmental risk, 97–98, 258–59
environmental sanitation, 196
Environmental Science & Technology, 21
EPA. *See* Environmental Protection Agency
EPs. *See* Equator Principles
Equator Banks, 229
Equator Principles (EPs), 97, 227–30, 256, 280n6
ERI. *See* Energy Research Institute
ESG. *See* Environmental, Social, and Governance
ETFs. *See* Exchange Traded Funds
ETS. *See* emissions trading system
EU. *See* European Union
European Commission, 235
European Investment Bank (EIB), 232

European Union (EU), 235, 245, 247, 249
Euro V standard, 47
Euro VI standard, 47
EVs. *See* electric vehicles
excessive consumption, 80
Exchange Traded Funds (ETFs), 231–32
Export-Import Bank of Korea, 233
export price equation, 123

factor market-clearing, 124
Fair Winds Charter, 158, 159, 273n17
Fannie Mac, 228
feed-in tariffs (FiTs), 236, 237–38
fees: coal and, 208, 209; congestion, 43, 70, 91, 93, 148–49, 191; parking, 191
financial policies, 78
financial resources, 163
First Trust NASDAQ Clean Edge Green Equity Fund, 232
fiscal balance, 102
fiscal expenditure, 110–11
fiscal policies, 78; coal and, 205–7
fiscal resources, 163
fiscal revenue, 90, 91, 110–11, 190, 194
fiscal spending, 88
FiTs. *See* feed-in tariffs
fixed roof tanks, 51
floating roof tanks, 51
flue gas denitration, 43
forced shutdowns, of coal boilers, 207
fossil fuels, 24. *See also specific types*
France, 64, 243
FTSE All-share Index, 244
FTSE Good Environmental Leaders Europe 40 Index, 233
FTSE Japan Green Chip 35 Index, 232
Fuel Standards: National III, 44, 45; National IV, 44, 45, 47, 66, 169; National V, 44, 47, 111

fuel standards upgrade, 44
fuel switching, 158–60
Fukushima accident, 67
Fullgoal Low Carbon Environmental Protection Equity Fund, 232
full reform, 202, 204–5, 210

Gallup, 81
Gaoqiao Petrochemical, 144
gasoline, 44, 47, 160
Gates, Bill, 226
GDP. *See* gross domestic product
GE. *See* General Electric
General Administration of Quality Supervision, Inspection and Quarantine, 45
General Electric (GE), 81, 196, 223
Generation Investment Management, 230
GENPACK software, 126, 128
German Renewable Energy Act, 238
Germany, 40, 41f, 42, 81, 87, 108–9, 235; green loans in, 236; solar power and, 155, 238
GIB. *See* Green Investment Bank
Global Carbon Index Fund, 232
Global Environment Fund, 230
Global Satellite Monitoring, 21
Global Warming Index (UBS-WGI), 232
goGreen car loans, 228
government agencies, 196
government deficits-exogenous equations, 121
government interventions, 16
government policies, 58–59; administrative, 101; CGE and, 103–5; coal, recommendations and, 207–10; financial, 78; fiscal, 78, 205–7; green finance, framework for, 18, 95–100; for incentives, 82–100; land, 78; Pareto-optimal economic, 94; shocks, 106; structural adjustments: green finance

framework, 95; in Shanghai, 145–50; top-down design process, 20
government spending, 121
government spending-exogenous equations, 121
Greater London Authority, 5
Greece, 238
green banks, 95–96, 234; government funding of, 239
Green Bond Catalog, 96
green bond market, 96–97, 257
green bonds, 232–34; interest rates and, 233; investors and, 233–34; management of proceeds, 233; purpose of, 233; tax exemption for, 238
green consumer products, 16–17; consumer preferences for, 263; public education and, 100
green credit cards, 228
green credit ratings, 242–44
green ETFs, 231–32
green finance: economic framework for: consumer problem and, 224–26; producer's problem and, 221–24; establishing system of, 252–63; green investment and: carbon trading system, 247–50; credit ratings and, 242–44; environmental factors of, 241–42; institutional investor networks, 246–47; introduction to, 220–21; lenders' environmental liability, 239–41; policy framework, 18, 95–100; public environmental impact assessment system, 250–51; public finance and, 236–39; types of products: EPs, 227–30; green banks, 234; green bonds, 232–34; green ETFs, 231–32; green insurance, 234–36; green loans, 227–30, 236; green private equity, 230–31; mutual funds, 231–32; venture capital funds, 230–31

green footprint, 262
green indices, 232
green infrastructure projects, 92
green institutional investor networks, 246–47
green insurance, 234–36; mandatory, 98–99, 260; US and, 235
green investment, 239–51
Green Investment Bank (GIB), 232, 234, 239
green investors network, 99, 263
green loans, 227–30; government provision of guarantees, 237; interest subsidies on, 236–37, 257–58
Greenpeace, 4, 21, 24, 166, 225
green private equity, 230–31
gross domestic product (GDP), 6, 20, 32, 35, 37, 51, 56–57, 102, 106; of Beijing, 196; coal and, 202; heavy industry and, 13; real growth, 166; secondary industry and, 54–55; services and, 143–44; Shanghai and, 132, 135
Guide to Best Practical Technologies for Pollution Prevention in the Cement Industry, 7
"Guiding Opinion on Launching Work on a Pilot Program of Mandatory Pollution Liability Insurance" (MEP and CIRC), 98
Guo Ju'e, 200, 201

Hahn, 118
Hann, Daniel, 242
Han Wenke, 8
Hawaii, 234
health hazards, 4–6
Healthymagination, 81
heavy industry, 14, 77, 115, 141, 185; in Beijing, 179, 196–97; GDP and, 13; manufacturing, 139; in Shanghai, 143–44; US and, 179

Hebei, 94, 95, 172, 173–75, 184–86; regional compensation for, 180–88

"Hebei Atmospheric Pollution Prevention and Control Target Responsibility Statement," 186

"A High Level of Importance Must be Attached to the PM2.5 Pollution Prevention and Control Problem" (Zhou), 7

high value-added services, 156–58

Hong Kong, 73, 92, 139, 152, 156–58

Hongyu Information, 67

HSBC Global Climate Change Benchmark Index Fund, 232

Huai River heating demarcation line, 4–5

Hu Min, 24

hybrid vessels, 159

hydrocarbons, 45

hydropower, 34, 36, 60, 104, 136; in Africa, 242; energy structure and, 68; in Shanghai, 155

ICAP. *See* International Carbon Action Partnership

IFC. *See* International Finance Corporation

IIGCC. *See* Institutional Investor Group of Climate Change

IMO. *See* International Maritime Organization

import-export trade modules, 122

import price equation, 123

incentives, 77–78; market distortions and, 78–82; market failures and, 79–80; market participants' problem and, 78–82; policies and actions for, 82–100; social responsibility and, 80–82; state intervention problem and, 78

income, per capita, 163

INCR. *See* Investor Network of Climate Risk

India, 230, 243

indirect adjustments, 153

indirect tax: price equation for, 123; rate, 103, 114

Indonesia, 225, 230

Industrial Bank Co. Ltd., 229

industrialization, 54, 215

industrial land prices, 82–83, 103, 216

industrial soot, 47, 48–49

industrial VOCs, 50–51

industry: coal and, 102, 213–19; emissions, 47–51; growth of, 33; manufacturing, 139, 141; nonmetal mining, 48; overconcentration of, 139–41; secondary, 54–55, 56, 174, 217–18; shipping, 142–43, 156–62; structural adjustments and, 53–59. *See also* heavy industry

inflation, 111–14, 149

initial public offerings (IPOs), 92, 231

Institute for Health Metrics and Evaluation, 5

Institutional Investor Group of Climate Change (IIGCC), 246

insurance: civil liability, 235; environmental liability, 234; green, 98–99, 234–36, 260

interest subsidies: on green loans, 236–37, 257–58; programs, 97

International Carbon Action Partnership (ICAP), 247

International Energy Agency, 66–67, 154

"International Experience with Promoting Green Credit: The Equator Principles and IFC Performance Standards and Guidelines" (MEP), 229

International Finance Corporation (IFC), 229, 232, 280n6

International Maritime Organization (IMO), 158, 159, 274n20

International Organization for Standardization (ISO), 244

International Shipping Center and Free Trade Zone, 142
Internet financial services, 133
interregional environmental compensation, 93–95
inventory demand, 121
Investor Network of Climate Risk (INCR), 246
IPOs. *See* initial public offerings
iShares Global Clean Energy ETF, 232
ISO. *See* International Organization for Standardization

Japan, 67, 142, 232, 243. *See also* Tokyo
Japanese Ministry of the Environment, 245
Japanese National Railways, 92
Jiao Jianling, 85
Jing-Jin-Ji. *See* Beijing-Tianjin-Hebei region
Jinshan Petrochemical, 144
Johannesburg Stock Exchange, 245
Johansen, Leif, 117
JP Morgan Environmental Index-Carbon β Fund, 232

KFC, 225
KfW Bankengruppe, 234, 236, 237
Klein-Rubin function, 121
Kong Xianli, 200
Kumar, R., 267n3

labor market-clearing, 124
labor price equation, 123
land: industrial prices of, 82–83, 103, 216; policies, 78; subway + land development model, 92
Leadership in Energy and Environmental Design (LEED), 228
lenders' environmental liability, 239–41
Leontief sets, 119, 127, 128, 210

leverage effect, of public finance, 236–39
LIBOR. *See* London Interbank Offer Rate
license plate auction systems, 43, 70, 88–91, 93, 145–48, 152, 189–91
"Light Automobile Pollutant Emissions Limits and Measurement Methods," 114
Li Hongbin, 4
Lin Boqiang, 9, 200, 201
Line 4, 192
liquid natural gas (LNG), 154, 159, 160
Li Shuping, 84
Liu Zixi, 8
LNG. *See* liquid natural gas
loans, green, 227–30, 236–37, 257–58
Lockyer, Bill, 233
London, England, 91, 133, 139, 149, 156, 157, 165
London Interbank Offer Rate (LIBOR), 233
Los Angeles, 133, 165
lottery system, 189–91
Luo Minghua, 84, 216
Lu Shize, 8
Lü Xinhua, 30

Ma Jun, 274n3
mandatory green insurance, 98–99, 260
manufacturing industry, 139, 141
Mao Zedong, 185
Marcazzan, Grazia M., 267n3
margin demand module, 122
Marine Environmental Insurance Law, 235
market: capital market-clearing, 125; distortions, 77, 78–82; equilibrium, 224; factor market-clearing, 124; failures, 77, 79–80; green bond, 96–97, 257; participants' problem, 77, 78–82; prices, 224; problem, 77

market-clearing module, 124–25
Market Stability Reserve (MSR), 249
Maryland Bank and Trust, 241
M&As. *See* mergers and acquisitions
Masih, Abul M. M., 200
Masih, Rumi, 200
Mass Transit Railway (MTR), 92
Mellon Bank, 241
membrane filter media dust collectors, 49
MEP. *See* Ministry of Environmental Protection
mergers and acquisitions (M&As), 197
Mexico, 230
microeconomics, 80, 222, 224
Middle East, 242
Miller, Brian G., 5
Ministry of Environmental Protection (MEP), 1, 45, 48–51, 98, 166, 172, 183, 229; denitration and, 42; desulfurization and, 41; official national PM2.5 target of, 29; reforms by, 3
Ministry of Finance (MoF), 183, 237
Ministry of Industry and Information Technology, 50
Ministry of Railways, 71
Ministry of Transportation, 159
MoF. *See* Ministry of Finance
Monash University, 103, 118
Montgomery, David W., 247
MSR. *See* Market Stability Reserve
MTR. *See* Mass Transit Railway
municipal bonds, 93, 152
Mun S. Ho, 9
mutual funds, 231–32

NASA. *See* National Aeronautics and Space Administration
NASDAQ Clean Edge US Index, 232
National Aeronautics and Space Administration (NASA), 21, 27

National Air Quality Software Company, 21
National Bureau of Statistics (NBS), 6, 163
National China V Fuel Economy Standard for Vehicles, 7
National Chung Hsing University, 93
National Cleantech Group, 227
National Development and Reform Commission (NDRC), 65, 68, 172, 260
National III Fuel Standard, 44, 45
National IV Emissions Standard, 66, 169
National IV Fuel Standard, 44, 45, 47
National V Emissions Standard, 44, 47
National V Fuel Standard, 44, 47, 111
natural capital liability, 5, 250, 251, 259
natural gas, 34, 36, 60, 64, 102, 136, 153; Beijing and, 178; energy structure and, 65–67; LNG, 154, 159, 160; proportion of, 195
Natural Gas Development, 65
NBS. *See* National Bureau of Statistics
NDRC. *See* National Development and Reform Commission
negative contribution, 15
negative externality, 79
Nestlé, 225
"A New Angle on Sovereign Credit Risk: Environmental Risk Integration in Sovereign Credit Analysis" (UNEP), 243
New Resource Bank, 228
"New Ventures" project, 230
New York, 133, 139, 151, 234
New York Stock Exchange (NYSE), 241, 251
New Zealand, 84, 216

Nielsen, Chris P., 9
NIKKO-FTSE Japan Green Chip
    Index Fund, 232
nitrogen oxides, 24
nitrous oxide (NOx), 27, 40, 42–43,
    45, 79, 86, 133; fees on, 209
noncoal emissions, 57
nonenergy commodity composites,
    128
nonmetal mining industry, 48
nonpower industries, 141
no reform, 202, 204, 210, 213
North Africa, 242
Norway, 117
*Notice Concerning the Effective
    Control of Urban Dust Pollution*
    (MEP), 51
NOx. *See* nitrous oxide
nuclear power, 36, 155; energy
    structure and, 67–68
NYSE. *See* New York Stock
    Exchange

OECD. *See* Organisation for
    Economic Co-operation and
    Development
offshore wind technology, 68
onshore power, 160–61
ordinary enforcement, 168, 172
Organisation for Economic
    Co-operation and Development
    (OECD), 6, 55, 132, 154, 257;
    automobile ownership and, 70;
    clean energy and, 20, 64; energy
    elasticity and, 54
ORNAIG model, 103, 118
"other" emissions, 268n18
ozone, 50

Pareto improvement, 188
Pareto-optimal economic policy, 94
Paris Bourse, 245
parking fees, 191
partial reform, 202, 204, 210
particulate matter (PM), 2, 43;
    secondary, 50; SPM, 133

passenger vehicles: increased use in,
    14–16. *See also* automobiles
Pearl River Delta (PRD), 3, 159,
    273n17
Peking University, 4, 24
Peng Yi, 208
petroleum products, 7, 33, 44, 135
photochemical smog, 50
Ping Xinqiao, 84, 216
planned environmental measures,
    170, 188
PM. *See* particulate matter
PM2.5, 1, 3, 15, 19–21, 93,
    268n9; 35 μg/m³ target for,
    29–30; Beijing: adapting control
    model for, 165–67; challenges
    in meeting target for, 172–79;
    coal and, trends under various
    scenarios, 204–5; creating
    model for, 30–34; data for,
    22, 23t; emissions reduction
    model for, 20, 27–34; energy
    consumption and, 33; estimates
    from control model, 38t, 39t;
    estimating effects of reduction
    and, 33–34; estimating levels
    of, 21–22; global emissions of,
    28f; international comparison of
    levels, 27; key assumptions for,
    32–33; levels in various global
    cities, 29f; negative contribution
    and, 15; primary sources of,
    22–27; recommended actions
    and, 34–39, 36t; results of model,
    34–39; in Shanghai: control
    model for, 134–36; structural
    impediments to meeting
    target in, 138–43; structural
    adjustments and, 35–39, 53,
    54f, 74–76, 161; technical
    environmental measures and, 35;
    transport volume growth and, 33;
    yellow-label vehicles and, 8
policy environment, 163
"polluter pays" principle, 147, 191,
    235

polluting products, 222
Pollution Control Laboratory, 24
population growth, 271n13
Portugal, 238
positive externalities, 79
power: consumer purchasing,
149; generation of, 153–54;
nonpower industries, 141;
nuclear, 36, 67–68, 155; onshore,
160–61; shore, 273n4. *See also*
hydropower; solar power; thermal
power; wind power
PPI. *See* producer price index
PPP. *See* public-private partnership
PRD. *See* Pearl River Delta
preferential taxes, 84, 216
PRI. *See* Principles for Responsible
Investment
price equation for margin and
indirect tax, 123
price module, 122–23
price subsidies, 237–38
pricing mechanisms, 58
Principles for Responsible
Investment (PRI), 241, 246
private automobile ownership, 22
private capital, 92, 95, 152
private equity, green, 230–31
*Proceedings of the National Academy
of Sciences of the United States of
America*, 4
Procter & Gamble, 225
producer price index (PPI), 102,
106
producer's problem, 221–24
production module, 119–21, 120f
profit maximization, 221–24
public acceptance, 136
public education, 100
public environmental impact
assessment, 98, 250–51
public finance, 221, 254; leverage
effect of, 236–39
"Public Finance Mechanisms
to Scale Up Private Sector

Investment in Climate Solutions"
(UNEP), 236
public opinion, 1, 30
public-private partnership (PPP), 92,
107, 192–93
public transportation, 104, 153,
196, 269n6; funding for, 194;
inadequacy of, 142. *See also* rail
transportation
public vehicles, 148, 191

Qiandao Lake, 182, 183
quantitative relationships, 30

Rabobank, 228
railroads, 15
rail transportation, 116, 136,
143, 271n11; in Beijing, 176;
financing system for, 91–93;
increased use of, 150–53;
infrastructure for, 71–72; mileage
per capita, 267n5; use of,
14–16, 34, 56. *See also* subway
transportation
Rankins CSR Ratings, 245
rebalancing, 13
reform: full reform, 202, 204–5,
210; by MEP, 3; no reform, 202,
204, 210, 213; partial reform,
202, 204, 210
regional compensation, for Beijing,
180–98; clean energy and, 194–
96; heavy industry and, 196–97;
for Jing-Jin-Ji, 184–88; subways
and, 188–94; in theory and
practice, 181–84; vehicle fleet size
and, 188–94
"Regulations Concerning
Environmental Protection in
Offshore Oil Exploration and
Development," 236
renewable energy, 154–55, 195. *See
also specific types*
Renewable Portfolio Standard
(RPS), 154–55

"Report on China's Green National Accounting Study" (SEPA), 6
reputation, 222
resource taxes, 85–86, 103, 104, 209–10, 212
retail green credit, 228
revenue: fiscal, 90, 91, 110–11, 190, 194; maximization, 121; new streams of, 194
road development, 15
road transportation, 20, 169
Royal Bank of Scotland, 251
RPS. *See* Renewable Portfolio Standard
Ruan Xiaodong, 7
Russia, 27, 195

seaborne shipping, 153
secondary industry, 54–55, 56, 174, 217–18
secondary PM, 50
Securities Law, 262
SEPA. *See* State Environmental Protection Administration
services: GDP and, 143–44; high value-added, 156–58; Internet financial, 133; tax burden on, 83–85
Seven Baselines, 83
Shaanxi-Beijing pipeline, 66
shale gas, 64, 66–67
Shandong Energy Group, 208
Shanghai, 4, 10, 18, 51, 86, 89–91, 93, 189; environmental actions in: effectiveness and, 137–38; existing and planned, 136–37; international experience and, 133–34; introduction to, 131–34; as leader, 132–33; PM2.5 control model for, 134–36; structural adjustments for, 131; clean energy, 153–56; heavy industry and, 143–44; policy options, 145–50; rail transportation, 150–53; services and, 143–44; shipping industry and, 156–62; subway travel and, 144–45; vehicle fleet growth and, 144–45; structural impediments and, 138–43; cars and, 141; clean energy and, 142; high pollution industry and, 142–43; low value-added shipping and, 142–43; overconcentration of industry, 139–41; public transportation and, 142
"Shanghai Clean Air Action Plan (2013–2017)," 136
Shanghai Environmental Monitoring Center, 135, 144
Shanghai Free Trade Zone (SHFTZ), 144, 157, 159
Shanghai Securities Exchange, 245
Shell China, 67
SHFTZ. *See* Shanghai Free Trade Zone
shipping industry, 142–43, 156–62
shocks, 106
shore power, 273n4
Sichuan-Eastern China pipeline, 66
Sinar Mas, 225
Singapore, 15, 73, 89, 139, 141, 146, 148, 156–57, 238
Sinopec, 111, 114
small- and medium-sized enterprises (SMEs), 230, 237
smog: atmospheric, 133; photochemical, 50
SO_2. *See* sulfur dioxide
social responsibility, 99–100, 222, 225, 226; CSR, 80–81, 223; incentives, 80–82
social stability, 147
social welfare, 90, 180, 188, 221–22
Société Générale, 242
solar power, 36, 60, 104, 155, 242, 270n6; energy structure and, 69; Germany and, 155, 238
soot, 27; control of, 33, 35; industrial, 47, 48–49
South Korea, 238

Spain, 238
S&P Global Clean Energy Index, 232
SPM. *See* suspended particulate matters
Standard Chartered, 228
Standardization Administration of China, 45, 47
Standard & Poor's, 243
Standing Committee, of Beijing Municipal People's Congress, 275n14
State Administration of Taxation, 217
State Council, 3, 86, 279n33; Development Research Center, 8
State Environmental Protection Administration (SEPA), 6, 7
state intervention problem, 77, 78
State Key Joint Laboratory of Environmental Simulation, 24
steel, 24, 48
Stockholm, 149
structural adjustments, 17, 30, 35–39, 103; automobile ownership and, 69–74; for Beijing, 180–98; coal and, 217–19; economic, 29; energy and, 55; energy structure and, 59–69; green finance policy framework and, 95; industry and, 53–59; PM2.5 reductions and, 74–76; scenario, 32; for Shanghai, 131, 143–62; transportation and, 56; types needed, 53–56
structural distortions, underlying causes of, 16–17
structural readjustment, 13
subsidies: on clean energy, 87–88, 110, 195, 210–13; in Hebei, 174; interest, on green loans, 236–37, 257–58; interest programs, 97; price, 237–38
subsidization formula, 147
substitution elasticity, 126

subway + land development model, 92
subway transportation, 15, 20, 34; capacity and, 193–94; expansion of, 72–74, 176; extending tracks for, 150–51; increased, 144–45; new infrastructure for, 151–53; regional compensation and, 188–94; ridership of, 191–93
sulfides, 24
sulfur dioxide (SO_2), 27, 40–44, 55, 79, 86, 133; fees on, 209
Sun Edison solar energy project, 230
Suning Corporation, 1
supply mechanisms, 58
suspended particulate matters (SPM), 133
Sustainable Finance Center, 229

Taiwan, 216
Tang Dengshan, 84, 216
taxes: carbon, 86–87, 104, 209–10, 212, 251; coal and, 205–7, 208; green bonds, exemption for, 238; indirect, 103, 114, 123; preferential, 84, 216; resource, 85–86, 103, 104, 209–10, 212; services, burden on, 83–85; VAT, 58, 84–85, 208, 216–17
technical environmental measures, 35
technical feasibility, 136
Technical Policy for the Prevention of Volatile Organic Compound (VOC) Pollution, 7, 50
Technical Specifications for Managing the Operation of Flue Gas Treatment Facilities at Thermal Power Plants, 7
Tendris Holdings, 228
TEUs. *See* 20-foot equivalent units
TFP. *See* total factor productivity
thermal power, 24, 48, 77, 202; coal and, 43; desulfurization of, 40–42

"Three-Year Action Plan for Environmental Protection and Construction," 136
Tokyo, 73, 82, 133, 139, 151, 165, 193, 216
top-down policy design process, 20
total factor productivity (TFP), 106
transportation, 37; elasticity, 33, 53; public, 104; road transportation, 20, 169; structural adjustments and, 56; urban structure of, 107; volume growth, 33, 55. *See also* public transportation
transshipment, 273n4
Trucost, 5, 244, 250, 252, 259
Tsinghua University, 4, 5, 7, 65
Turco Coatings, Inc., 241
Turkey, 243
20-foot equivalent units (TEUs), 143
"2013–2017 Work Plan for Accelerated Reduction of Coal Burning and Development of Clean Energy," 168, 178
2011 Annual Environmental Statistics Report (MEP), 48

UBS-WGI. *See* Global Warming Index
UN. *See* United Nations
UNEP, 236, 243
United Kingdom, 82, 108, 216, 238, 244, 250; green banks in, 234; green loans in, 237; heavy industry and, 179; natural gas and, 154
United Nations (UN), 241, 246
United States (US), 67, 81, 87, 142, 217, 232, 238; automobile ownership and, 70; green banks in, 234; green insurance and, 235; heavy industry and, 179; lender liability in, 241; real GDP and, 55
United States v. Mirabile, 241

urban commuting, 20, 139
urbanization, 15, 73
urban light rail, 72
urban transportation structure, 107
US. *See* United States

value-added tax (VAT), 58, 84–85, 208, 216–17
value of statistical life (VSL), 5
values-based consumers, 81
VAT. *See* value-added tax
vehicle fleet size, 141, 144–45, 175, 188–94, 270n6
venture capital funds, 230–31
Vestas, 226
Visa Greencard, 228
volatile organic compounds (VOCs), 19, 27, 47, 168, 170, 186; industrial, 50–51
VSL. *See* value of statistical life

Walrasian equilibrium, 102, 117
wastewater, 182–83
waterborne anticorrosive coatings, 50
water-to-water transshipment, 143
weighted average price, 127
Wei Weixian, 200
welfare, social, 90, 180, 188, 221–22
Wells Fargo Bank, 227, 228
West-East pipeline, 66
Westinghouse, 67
West-to-East Gas Pipeline, 154
WHO. *See* World Health Organization
willingness-to-pay (WTP), 5
wind power, 34, 36, 60, 102, 104, 116, 136, 242; in Denmark, 226; energy structure and, 68; Shanghai and, 155
World Bank, 5, 6, 163, 229, 232–34, 256, 279n33, 280n6
World Health Organization (WHO), 1; Air Quality Guidelines of, 2, 30

World Resources Institute (WRI), 230

WTP. *See* willingness-to-pay

Wusong International Cruise Terminal, 161

Xie Peng, 4

Xinanjiang Watershed environmental compensation mechanism pilot program, 182–84

X-trackers S&P US Carbon Emissions Reduction Fund, 232

Xu Xiaoliang, 200, 201

Yamanote line, 151

Yang Guang, 7

Yangshan Harbor, 161

Yangtze River Delta (YRD), 3, 182

yellow-label vehicles, 8, 135

Yin Aizhen, 200

Yitai Group, 208

YRD. *See* Yangtze River Delta

ZERO2IPO, 230

Zhai Qing, 268n8

Zhang Bo, 208

Zhang Chanjuan, 7

Zhang Lei, 200

Zhang Lunjun, 84

Zhang Xiliang, 270n6

Zhejiang People's Congress, 183

Zhonghai Environmental Protection and New Energy Fund, 232

Zhou Hongchun, 7

Zhou Quanlin, 84